BYE-BYE CARBON!
TOO BAD, SO SAD.

The Transformation of Automotive Thermodynamics

What does this book do? It tells car makers how to produce cars that can be used sustainably with optimized engines and transmissions. Fuel use is reduced to the actual ideal, resulting in massive fuel use and carbon footprint reductions. As of this publication, all automotive vehicles are obsolete and need to be replaced.

David Ronald Dick

Olympus Story House

TABLE OF CONTENTS

FORWARD

What does this book do? It tells car makers how to produce cars that can be used sustainably with optimized engines and transmissions. Fuel use is reduced to the actual ideal, resulting in massive fuel use and carbon footprint reductions.

The purpose of this paper is to reduce automotive fuel consumption without compromising transportation services. The objective is to reduce fuel use of the prime mover, either engine or electric motor, to closely match the theoretical energy needed to move the vehicle, regardless of its speed. Alignment of engine energy use with the actual energy needed to move the mass of the vehicle, all vehicles, at all speeds is the objective. This is not the case with automotive fuel use today, and if alignment of fuel use with energy need is achieved, the potential fuel savings are enormous. Carbon discharge to the environment would be reduced in the order of millions of tons per day.

Sustainability of our individualized automotive expectations is another motivation for doing so. Moving away from our present fuel misuse to a more compliant fuel use is another motivator. Doing so would better manage our hydrocarbon and electric supply resources. Oil and electricity supplies would last longer. Also, optimum fuel use would reduce the loading on waste resources. Carbon loading on the atmosphere would be less. Radiated heat from automobile operations would be less. These changes would be very beneficial in many ways.

But transformation can only take place if there is room for improvement over and above an unacceptable existing condition, and if a better position is available. The unacceptable condition of the automotive industry today is the speed dependent changing of fuel use and efficiency of all vehicles. This means that vehicle fuel efficiency varies with the speed of the vehicle. This is in non-compliance with the mechanical laws of energy. In the better position, optimum automotive fuel use would use energy at theoretically correct levels at all speeds and misuse would be eliminated. This means that fuel use, energy use, would vary directly with the work done in moving the vehicle. This "correct" automotive fuel use would be defined by existing mechanical laws of energy. The transformation would move automotive fuel use from present non-compliance to compliance with the laws of energy.

Thermodynamically, the automotive industry today, is in systemic non-compliance with mechanical laws of energy. The information for thermodynamic compliance has never been available to automotive system managers. Therefore, the industry could not have developed in any other way than it has. If new physics and technology was supplied to the industry, the transformation to optimum energy use, could and would take place. That new physics and technology is the purpose of this paper. And if it is correct, relevant and useful, it means that, thermodynamically, every vehicle on earth, Formula One included, is an obsolete relic of this bygone age and needs to be replaced. Imagine, replacing the present 1.5 billion vehicles, with energy compliant vehicles, and by doing so, reducing automotive fuel consumption by half, or more, equal to at least 750 million vehicles. The figure of 750 million is not an exaggeration and is a reduction that is achievable.

CREDENTIALS

The author, David Dick, holds a Bachelor of Theology degree, BTh, from Canadian Mennonite University in Winnipeg, Manitoba, Canada. His certification as a First-Class Stationary Engineer, Standardized, Ontario, Canada, has provided extensive experience with theoretical and practical thermodynamics. He recently retired as the lead pasteurizer at Hewitt's Dairy in Hagersville, Ontario. This position was a daily exercise in compliance to protocols and spatial awareness and logic. All three experiences have been essential in the research for this paper.

INTRODUCTION

This discussion presents a thesis for the transformation from present automotive fuel use standards to theoretically correct specifications, in five parts. The problem with present day automotive fuel use is first. Secondly, the science, the new science, that allows the transition away from the present, toward the new, follows. This science leads to the technology, the transmission, that solves the compliance problem, is in section three. The surprising, unexpected benefits arising from the technology come in part four of the paper. Here we discuss the plusses of full inertial braking and acceleration, and the optimization of internal combustion engines. Also included are larger possible systemic impacts related to the science. The conclusion follows part four.

Understanding the problem with automotive fuel use is essential before any development is possible. In part one, graphs will illustrate present day auto fuel use, ideal auto fuel use and the contrast between the two. In the context of the laws of thermodynamics, the meaning of fuel use, as per the graphs will be discussed. Then, with the application of the mechanical law energy, fuel use and misuse efficiency are discussed. Graphs will illustrate present auto fuel efficiency, ideal efficiency and the potential improvement from one to the other. A very clear statement of the essence of the auto fuel misuse problem finishes the first part of the paper.

The understanding of the auto fuel misuse problem sets the stage for the transition from the old to the new in part two. Due to the profound similarity of all auto fuel misuse in all vehicles, it becomes obvious that the science and technology for compliant energy use by automobiles does not exist. If any science that could enhance or enable compliant auto fuel use to take place was in existence, there is no doubt, it would be in use today. Therefore, this transformation paper must include a scientific investigation that develops new physics. The resulting fourth and fifth laws of motion combined with Newton's first three, provide the context for solving the automotive fuel use non-compliance problem in section two.

The new physics sets the stage for developing the technology for solving the fuel use noncompliance in section three. Section three transforms the invisible into the visible. The new technology, a transmission, must translate the theory of the new physics into the real world. The discussion of the transmission will outline the process by which energy from a source is applied to a load in compliance with the laws of energy. This obedience to the laws is maintained by the transmission regardless of the force applied to the load, or, of the speed ratio between the engine and the load. This new physics compliant transmission solves the auto non-compliance fuel use problem.

The new technology of the transmission, revealed, two unexpected and surprising potential benefits of great importance to automotive transportation. This is section four. Initially, the machine was designed solely to achieve automotive fuel use compliance to energy laws. But the ratio range capability of the new transmission defines any and all speed ratios between the prime mover and the load. This ontological, energy law compliant, ratio capability of the transmission allows for the development of two very useful automotive applications. Firstly, the engine runs at steady speed and, as a result, can be fully optimized.

Engine energy efficiency has the potential of being doubled or tripled to fifty or sixty percent!

Secondly, full inertial braking is possible. This means that the braking process stores vehicle inertia to be used later for acceleration. All braking and acceleration would be flywheel based and inertial energy of the vehicle would be stored in a flywheel for braking. Flywheel energy would be recycled to the vehicle for acceleration. These developments would greatly improve fuel use efficiencies during speed transitions, up or down, or at steady speed conditions with highly efficient optimized engines. Engine optimization and full inertial braking and acceleration would greatly enhance automotive transportation efficiency and are achievable with this new science-based technology. Environmental sustainability may potentially reach net zero balance.

The conclusion paints a picture of where we could be going with our energy compliant automobiles. Some speculation on the far-reaching implications of the new science and technology shows the potential impact on society. The importance of this book and its potential positive impact on society is difficult to over-state.

PART ONE

THE PROBLEM

INTRODUCTION

The purpose of this first part is to find a clear understanding of the automotive fuel misuse problem. Graph A shows automotive fuel use as it is today. Graphs B and C show compliant fuel use and the possible improvement in energy use management. The meaning of this fuel use is analyzed in the context of the laws of thermodynamics.

Then Graph D shows automotive fuel efficiency. This graph is the result of the application of the mechanical laws of energy to Graph A, the fuel use graph. Graph E will show the ideal, compliant fuel efficiency. Next, Graph F illustrates the improvement in efficiency from present standards to energy compliant fuel use.

Analysis of the cause of the automotive fuel misuse follows. With the help of the laws of thermodynamics, the causes of the transformation of fuel energy to low grade heat are discussed. Understanding these non-compliant energy downgrades to low grade heat, sets the stage for the transition section, the science section, that follows and theoretically solves today's fuel noncompliance problem.

WASTEFUL FUEL USE.
GRAPH A

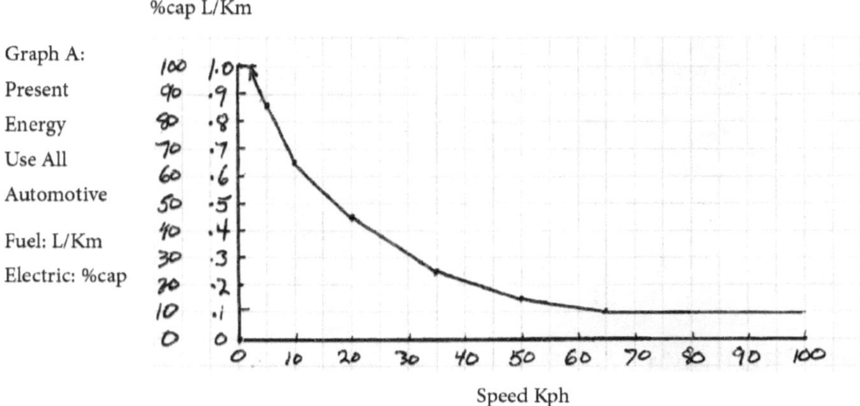

Graph A is the starting point of the discussion of the problem of automotive fuel misuse. This is the fuel use curve, the energy use curve, of every land vehicle on earth. Everything from the smallest mini bike to the largest diesel-electric locomotive uses fuel according to this curve! The graph is laid out with the usual horizontal and vertical axes. The horizontal axis represents vehicle speed in kilometers per hour (kph). The speed range of 0 kph to 100 kph was chosen for the scale of the horizontal axis to most closely represent real world vehicle speed. 0 kph, or the slowest speed is located at the left end of the axis and the highest speed is at the right end of the axis. The vertical axis displays vehicle fuel use rate. Two energy use scales, electric and gasoline, were chosen for the vertical axis. Minimum gas or electric use is located at the bottom of the axis and maximum energy use is at the top of the vertical axis. The electric scale range runs from 0% capacity minimum to 100% capacity at maximum. The gasoline fuel use varies from 0 L/K at minimum to 1.0 L/K at max.

The fuel use or energy use numbers of the vertical axis are different for the various machine sizes. Smaller vehicles would use less energy and larger machines would use more energy through the speed range. But the shape of the curve of Graph A remains the same for all vehicle sizes. The energy use numbers on this graph are for a typical mid-sized American built, four door sedan.

So, what is this graph showing us and what does it mean to say? This graph compares vehicle fuel use with vehicle speed. The curve shows us that fuel use of vehicles is at maximum at the beginning of acceleration from stop, essentially at the slowest speed. The fuel use rate at the beginning of acceleration is 1L/K, or equal to 100 liters per 100 kilometers. Then as the speed of the vehicle increases past say 20 or 30 kph, the fuel use drops from max at 1 L/K down to .4 to .5 L/K. Then as the vehicle reaches the highway speed of 80 kph the fuel use drops to .1 L/K or about 10 liters per 100 kilometers, or one tenth of max as it was at the beginning slowest speed! This is down from 100 liters per 100 kilometers at the beginning of the acceleration!

Note that fuel use is at max at slowest speed acceleration and at minimum when vehicle is at highway steady speed. One would expect fuel use to vary directly with vehicle speed. One would expect that as vehicle speed increases, more work needs to be done by the drive line to maintain speed and fuel use would therefore be higher. But Graph A above does not show this. It shows the exact opposite.

It shows us that automotive fuel use is at maximum at slowest vehicle speeds. To accelerate a vehicle from a standing start, a force has to be applied to it, according to Newton's First law of Motion. The engine applies this force by using mechanical power and energy from the prime mover to create friction with the clutch as the force is applied to the vehicle driveline and move the vehicle. At the moment of application of force, before the vehicle has started to move, all engine fuel use and engine mechanical energy is used as friction to generate the drive forces needed to accelerate the vehicle. The vehicle has not yet started to move and has therefore done no mechanical work but the engine has used fuel energy and mechanical energy in the application of force to the load, the standing vehicle. And in so doing it is using the maximum volume of fuel when, theoretically it should be using no fuel in the application of the drive force. This, as defined by mechanical laws of energy, is "zero efficiency" energy use at highest fuel volume, merely to apply a drive force onto the drive line!

According to the work done by the engine, this is a total waste of fuel!

Every vehicle, as it applies power to begin acceleration, but before it has started to move, uses energy at zero efficiency. The clearest manifestation of Graph A, and the "incorrect" fuel use in the real world is the American Top Fuel dragster race car. It's a race car designed to race for a quarter of a mile from a standing start, in the shortest possible time, to win the race over a competitor racing the same "drag." The driveline of such a race car consists of an engine driving a large multi-disc clutch that then drives the wheels through the rear transfer case. Engine speed, wheel speed and transfer case gearing have been calculated so that all speed change for the acceleration of the race takes place by the increasing engagement of the multi-disc clutch. The operating ratio range of the clutch is from full slip, or 1 to 0 to full engagement or 1 to 1 ratio. A typical race applies full engine power to the wheels via the engagement of the clutch. At the start of the race, full engine power, some 10,000 hp, is applied to the clutch at the full slip ratio of 1 to 0. At the point of engagement, all of the mechanical energy of 10,000 hp is converted to heat with friction in order to apply the accelerating force to the drive wheels. This friction takes place at the multi-disc clutch due to the full differential speed ratio between the engine or driver discs and the wheels and driven discs in the clutch. The driver discs bring full engine power to the clutch and move at engine speed. The driven clutch discs are connected to the rear drive wheels via the transaxle according to wheel speed. The driven discs are at "zero" speed at the start line. At the point of clutch engagement, full differential disc speed ensures maximum friction and thereby maximum drive force to the rear wheels. Then as the race car begins to accelerate, due to the applied frictional clutch force, the relative differential speed ratio of the drive and driven clutch discs becomes less. Half way through the race, the slip ratio of the clutch is about 1 to 1/2, soon to be fully engaged at 1 to 1, with no slip or relative speed differential at the end of the race. At the finish line, the clutch engagement is complete, clutch disc speed ratio is at 1 to 1, and friction-based force applications to the load have ended. Drive line mechanical energy efficiency is at maximum, when dragster speed is at max, but thermodynamic engine losses remain unchanged at 75% throughout the race. As the race car accelerates, the clutch slips less and less until it reaches the 1 to 1, non-slip ratio, at the finish line, when the race is over, 4 seconds later and at more than 300 mph! This is the epitome of American automotive muscle!

The reverse efficiency of mechanical energy use, by the multi-disc race car clutch, illustrates present day automotive fuel use and misuse. At force application but before motion has started, all the mechanical energy from the engine is downgraded to heat via friction, in order to apply a force. Automotive driveline systems today, use prime movers and transmissions with multiple gear ratio ranges, not multi-disc clutches, but the fuel use characteristics are the same as graph A. However, automotive fuel use during acceleration is not friction based as in the dragster. Slow speed, high mechanical advantage ratios in the lower gears of the automotive transmission use fuel at low efficiency and according to the same curve. This "zero" efficiency application of force is in complete non-compliance with the mechanical law of motion: "Work equals Force times Distance." Graph B shows what compliance to the work equation looks like.

But further to the problem, and outside of the scope of Graph A is vehicle braking. All vehicles at speed need the capacity to be slowed to a stop by braking forces. The braking systems of almost all vehicles slow the vehicles using friction-based disc or drum brakes. The design of the brakes is such that the disc is attached to the wheels and spins 1 to 1, the same speed as the wheels. Frame mounted friction pads apply force onto the discs when braking forces are needed to slow or stop the vehicle. The applied force of the shoes onto the discs creates friction. This friction creates the heat needed to reduce the energy of inertia of the moving vehicle to "zero" or full stop, if needed. The mechanical energy of the inertia of the vehicle is downgraded to low grade heat as is required by the first law of thermodynamics for the braking to be possible. Note that this braking process results in the reduction of the high-grade energy of vehicle inertia to low grade heat, radiated to the atmosphere, never again to be to be used as high-grade mechanical energy for vehicle motion.

What do Graph A and "braking" demonstrate about automotive use of energy? Graph A, the fuel use curve, shows us that the acceleration of vehicles from standing to highway speed is done with the maximum fuel use possible. Then at steady-state highway speed, mechanical power efficiency reaches some 85%. This is respectable, but the mechanical power is produced by the engine at thermodynamic losses of 70 to 80%. The mechanical engine power that is being used at a steady speed 85% efficiency is being produced at 20 to 30% efficiency by the engine!

The inertial energy of all moving vehicles today is being achieved at the highest possible fuel use imaginable!

Then, this costly, precious inertial energy is down graded to low grade heat by frictional braking forces if the vehicles need to stop. Then, acceleration back to steady highway speed uses fuel at highest volume and lowest efficiency according to graph A, and in non-compliance with the laws of energy.

Graph A is the fuel use curve for the acceleration of a vehicle from a standing start when energy is used "incorrectly" and in non-compliance with the mechanical law of energy, W=F X D, or Work equals Force times Distance. This simple equation defines the "correct" use of automotive energy and graph A shows how incorrect today's automotive fuel use really is.

It is difficult to imagine any machine with a more extravagant thirst for fuel than the automobile!

All transportation requirements are met by vehicles today, but at the cost of fuel use at the lowest possible efficiency with the highest possible consumption. When compared to the mechanical energy needed to move a vehicle as defined by "Work equals Force times Distance", present day fuel use is much higher than it should be. Graph B explains.

GRAPH B

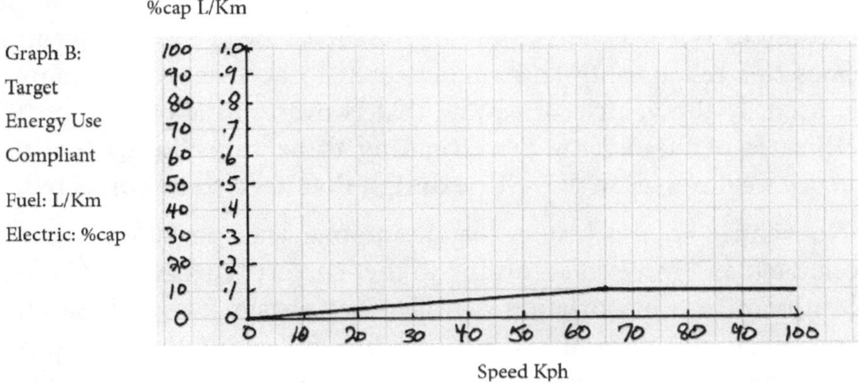

Graph B is the energy use curve for the acceleration and motion of a vehicle from a standing start. When energy is used "correctly" and in compliance with the mechanical law of energy, W=F X D, or Work equals Force times Distance, fuel use follows graph B. The simple three letter equation opens the door for the appropriate use of energy, the correct or compliant use of energy, for the acceleration and motion of a vehicle, potentially of all vehicles.

The horizontal and vertical axes of Graph B are identical to Graph A, so that comparisons between non-compliant and compliant automotive fuel use are clear and obvious. The horizontal or "x" axis represents vehicle speed. The scale of the vehicle speed from 0 to 100 kilometers per hour was chosen to most closely represent the speed of a typical vehicle on public highways. The vertical or "y" axis represents fuel use or energy use of a vehicle accelerating from a standing start to road speed, or sustainable, steady motion. The electric energy or fuel use scales of the vertical axis are meant to represent energy required to accelerate and hold speed for a typical American made four-door sedan.

Graph B compares the fuel use of a vehicle in motion when the supply energy equals the energy needed to move the vehicle as defined by W=FXD, the mechanical law of energy. All superfluous drive-line energy requirements have been eliminated. Graph B could be called the "ideal" automotive fuel use curve. The mechanical law of motion correctly defines the energy required to accelerate and hold vehicle motion and the vehicle fuel use should deliver the same, correctly defined energy to do so. But automotive fuel use of all vehicles today, follows Graph A, not Graph B.

Theoretically, thermodynamically, no vehicle has ever used more energy than Graph B for the acceleration and motion of the mass of the vehicle. The only amount of energy extracted by a vehicle from its drive-line to accelerate and travel has always only been the amount needed according to W=FXD and Graph B. The First Law of Thermodynamics defines this relationship and guarantees it to be true. Any amount of fuel used by the vehicle drive-line that is over and above Graph B is the misuse of fuel energy. **Investigating these superfluous fuel uses exposes weaknesses in drive-line design that need to be corrected.**

No vehicle in the history of automobile transportation has ever accelerated and traveled according to the "correct" energy use of Graph B. Automotive transportation fuel flow has always followed Graph A. The essence of automotive drive-line fuel misuse is the engagement of the non-moving, stationary load with a moving power source, for the application of force to start the acceleration of the vehicle. This differential speed between power and load, simply to apply a force to move the vehicle, downgrades all the mechanical energy to low grade heat, just to apply a force. This energy should have been used for vehicle motion, instead of the application of the drive force, which theoretically requires no energy to apply. All automotive drive-line designs work this way and use fuel according to graph A. The science and technology needed to move automotive fuel use to Graph B is developed later in this paper. The benefits of such developments, follows, in Graph C.

WASTED FUEL

GRAPH C

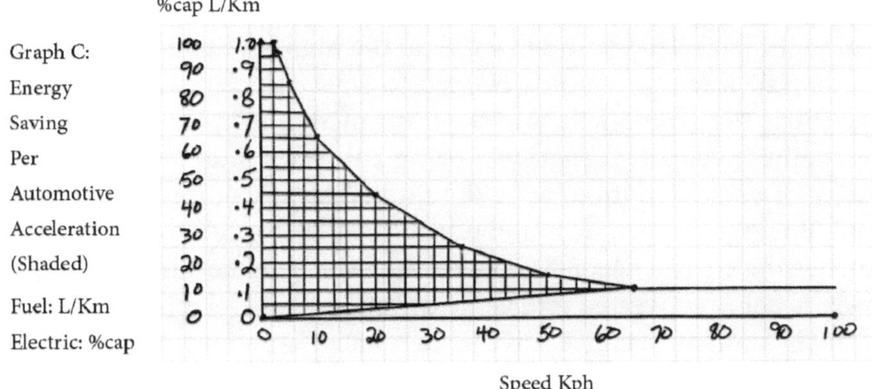

Graph C is the comparison of Graph A and Graph B. It compares actual automotive fuel use with theoretically "correct" fuel use. The cross-hatched region between the curves shows the fuel use savings for each automotive acceleration if automotive drive-lines used fuel "correctly" or in compliance with the mechanical laws of energy. This cross-hatched region of graph C defines the mechanical energy from the vehicle power source that is downgraded to low grade heat in order to create a drive force to accelerate the vehicle. The higher fuel use is required by the drive line that needs slow speed transmission gearing to supply the engine with the higher mechanical advantage needed to overcome the mass inertia of the vehicle to accelerate it. And according to the laws of energy, no energy is required for the application of a force. Therefore, fuel used to apply a force is used unnecessarily, or misused, wasted, and the crass-hatched section of Graph C shows this.

Graph C is designed in the same way as Graphs A and B. The horizontal axis represents vehicle speed with a speed scale typical for public highways. The vertical axis represents energy use, both hydrocarbon and electric. The scale for energy use is for a typical sedan on public roads.

Graph C very explicitly shows the fuel savings potential, if drive-lines could use fuel according to the mechanical laws of motion. At zero vehicle speed, and at the moment of force application to start acceleration of the vehicle, fuel use, misuse is at its highest volume and at zero efficiency! This is 100% wasted fuel use. The fuel use savings potential at this point is the highest. Then as the vehicle begins to accelerate, some of the engine power is being converted to the mechanical energy of vehicle motion and less to the drive force application. Graph C shows this with a decreasing curve. When the vehicle has accelerated to 20 kph, the curve of graph C has dropped to a fuel use of .45 L/Km, less than half of the starting point consumption of 1 L/Km. This means that more of the engine mechanical energy is being properly converted into the mechanical energy of vehicle motion. Efficiencies are still low due to the high mechanical advantage, low displacement gear ratios needed to start vehicle motion. Then as the vehicle reaches 65 kph, and the transmission is in "high gear," all of the engine mechanical energy is being converted into the mechanical energy of vehicle motion, as it should be. At this point, graph A and graph B merge, as seen on graph C, and the fuel use curve decreases and matches the energy needed to power the vehicle. Here, all of the engine power is being converted into vehicle motion, and fuel misuse solely for the application of a vehicle drive force for acceleration is eliminated.

But the fuel use of all vehicles follows Graphs A, and C, due to the mechanical design and operating processes of all drive-line systems. Any minor alterations or enhancements of existing drive systems cannot access theoretically correct and compliant fuel use in the drive of automotive vehicles. Only new science and technology, presented later in this paper, can initiate and manage such a massive transformation.

Graphs A, B and C, with the focus on fuel flow, very explicitly show the potential fuel use savings if automotive drive-line designs could operate in compliance with energy laws. The following graphs, D, E and F, analyze automotive fuel use mathematically, and generate efficiency curves that reveal implicit potential fuel use savings as clearly as the flow graphs do. These graphs follow.

VARYING FUEL EFFICIENCY
GRAPH D

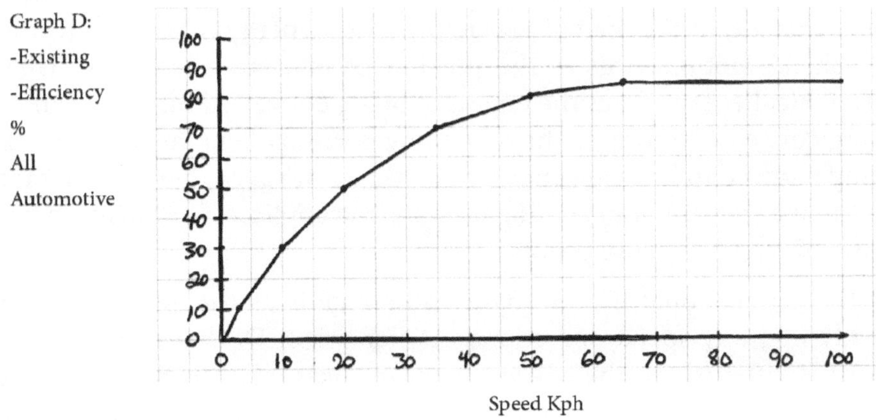

This is the fuel use efficiency curve of all automotive transportation. It displays the fuel use efficiency of all vehicles, both hydrocarbon and electric powered. The horizontal axis of the graph represents vehicle speed and the scale from 0 to 100 is in kilometers per hour, KPH, typical for highway speeds. The vertical axis represents vehicle fuel use or energy use efficiency. The scale of the efficiency axis starts at 0% at the bottom of the axis and rises to 100% at the top. The units of the efficiency scale are "percent", %, a unit calculated mathematically.

The efficiency calculations to generate Graph D focus on the use of mechanical engine power in the propulsion of a vehicle. Engine power is the only supply energy included in the efficiency calculations. The calculation compares engine power produced with actual energy used by the moving vehicle. Engine or electric motor power measured in kilowatts is compared to vehicle motion as it is defined by W=FXD, work equals force times distance, the mechanical law of energy. This comparison of mechanical engine power produced with actual energy used for the motion of the vehicle, at all speeds from 0 to 100 KPH, is used to generate the curve for Graph D.

The efficiency of fuel use varies relative to vehicle speed. At the beginning of vehicle acceleration and at the moment of engine power application of drive force, before motion has begun, engine power energy use efficiency is 0%. This is at the beginning of the curve where both efficiency and vehicle speed are 0 % and 0 KPH. This means that all of the engine power, usually very high at the beginning of acceleration, at the point of engagement of force application to the load is being down graded to low grade heat in order to apply the force. This, by definition of W=FXD, is zero, 0% efficiency, since the displacement D is zero. Then as the vehicle starts to move, actual mechanical work is being done by the engine power. As per Graph D, at 20 KPH, drive-line efficiency is 50%. This means that half of the engine power energy is being converted to vehicle motion. This is the portion of efficient energy use when mechanical energy from the engine is being converted into mechanical energy of vehicle motion. The remaining portion of engine power is still being reduced to low efficiency through the low speed, high mechanical advantage transmission gearing used for the application of drive forces, a low efficiency fuel use process.

The proportional use of engine power ends as the vehicle reaches steady highway speed operations. As per graph D, this takes place at 65 KPH, after which the efficiency curve levels off. This means that all engine power is now used at maximum drive-line efficiency at moving the vehicle. The misuse of fuel for the production of drive forces for vehicle acceleration is ended. Any engine power use at steady highway speed above 65 KPH, will be used according to the level efficiency curve, and engine power will be used to propel the vehicle and for vehicle motion only. For this reason, the efficiency curve levels off at 85% above 65KPH.

The misuse of engine power for the generation of a drive force for vehicle acceleration is shown by the curved part of Graph D. By definition, the application of a force requires no power or energy from the source according to W=FXD. Yet every acceleration of every vehicle on earth transforms energy into heat so that an accelerating drive force can be generated. What would Graph D look like if the drive force of acceleration could be generated without the misuse of engine power and in compliance with the laws? Graph E follows.

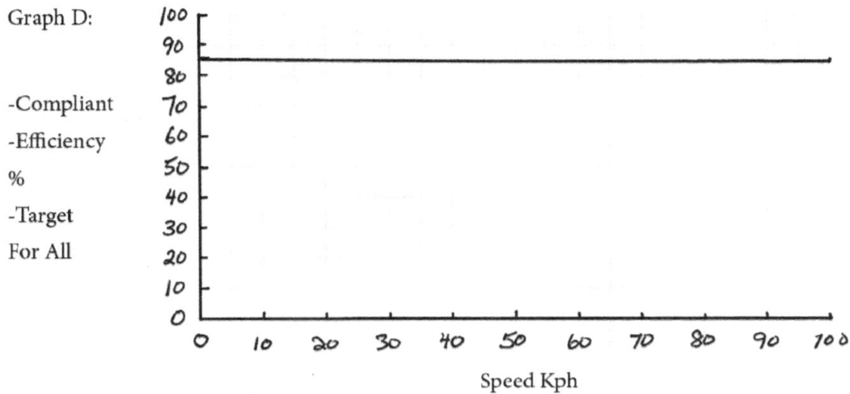

Graph E represents the efficiency curve of the acceleration and motion of a vehicle, when engine power is used in compliance with the mechanical laws of energy. No misuse of power for the generation of a force is present on this graph. If vehicle acceleration and motion could take place in compliance with W=FXD, energy use efficiency of the vehicle would look like Graph E, above. The horizontal axis represents vehicle speed with a scale of 0 to 100 KPH typical of public roads. The vertical axis represents percent efficiency from 0% at the bottom to 100% at the top of the scale.

The graph was drawn as a straight line at 85%. This is lower than 100% due to the internal loses of automotive drive-line systems. Engines and transmissions operate in an oil bath that provides the necessary lubrication needed for meshing metal parts, but viscosity and residual friction increase internal drag, both of which require energy from the power source to operate. These internal drive-line loses usually take about 15% of the available energy to overcome, reducing available efficiency to the 85% represented in Graph E.

There is no curved section of this graph like in Graph D. The curved section in Graph D was the result of energy from the power source being converted to engine waste heat to generate a force needed to accelerate the vehicle. This, by definition, is a low efficiency use of engine power, therefore, the curve. But Graph E presupposes the compliant use of energy by the drive-line to accelerate the vehicle. This means that the drive forces needed to accelerate the mass of the vehicle were generated empirically, and in compliance with the laws of energy. With the compliant force generation available for vehicle acceleration, all of the engine power is used for vehicle motion only, and can therefore be used at 85% efficiency throughout the speed range of the vehicle. This results in the straight line for the graph with no curve, like in Graph D. The comparison of Graph D and Graph F follows.

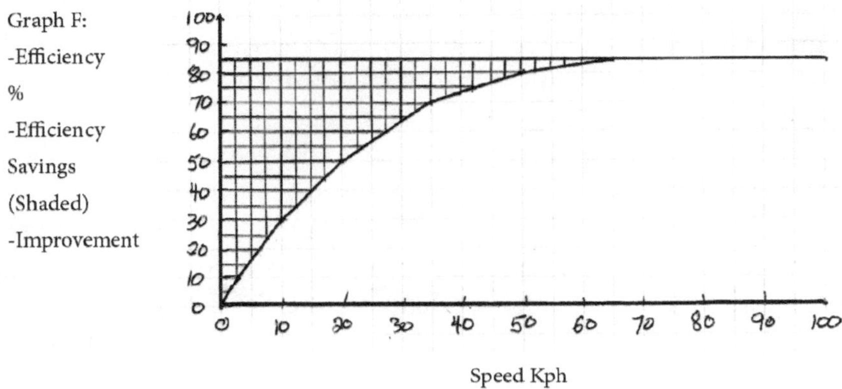

Speed Kph

Graph F is something special. The cross-hatched section of the graph shows the efficiency improvement when comparing compliant with non-compliant automotive fuel use. Graph F is Graphs D and E placed on the same set of axes, and the area between them highlighted. The horizontal axis represents vehicle speed in kilometers per hour, KPH, with a scale from 0 to 100, speeds typical for public highways. The vertical scale represents efficiency in terms of percent, %, with a scale of 0 to 100, all inclusive for an efficiency range. Both speed and efficiency scales start at 0 at the intersection of the axes and increase as they move outward.

The cross-hatched region between the graphs shows the potential automotive fuel use, energy use, efficiency improvement when moving from present to compliant. Compliant is theoretically correct. The upper straight-line graph shows energy laws compliant fuel efficiency, and the lower curved graph shows present day, non-compliant, automotive fuel efficiency. As the highlighting demonstrates, the greatest improvement in efficiency is possible at 0 vehicle speed. This takes place at the beginning of vehicle acceleration when engine power applies drive forces, but vehicle acceleration has not yet begun. At the moment of engine power engagement for acceleration, all engine energy is converted to heat to generate the needed acceleration force. This energy use is defined as 0% efficiency and the graph is at 0%. When this is compared to a drive-line capable of empirical, non energy-based force efficiency of 85%, the straight-line graph, the efficiency improvement is the largest. Then, as the vehicle begins to move, the efficiency improvement becomes less as more engine energy is being used for vehicle motion, as it should be.

As the vehicle begins to move, the portion of engine energy used for force generation at 0% efficiency becomes less and the portion of engine power used for the mechanical energy of vehicle motion increases. Engine energy use efficiency increases, rising above 0%, on its way to 85% at steady highway speed. At a speed of 20 KPH, the efficiency of engine energy converted to vehicle motion is 50%. This means that the remaining 50% of engine energy is still at low efficiency due to downgrading from low displacement, high mechanical advantage transmission gearing for the acceleration phase of the vehicle. Then as the vehicle speed reaches 65 KPH, the low efficiency force generation process by the engine ends, and all power is completely converted to vehicle motion at 85% efficiency, down 15% due to unavoidable internal driveline loses as per Graph E. Compliant fuel use for the acceleration and drive of a vehicle never uses or wastes engine power for force generation only, and always uses engine energy for vehicle motion at 85% thermal efficiency throughout the entire vehicle speed range.

Graphs A to E illustrate the problem with automotive fuel use. Obvious also is the cause of the energy misuse of all automotive transportation. The following summary of this problem will define the causes of non-compliant automotive fuel use.

SUMMARY OF THE PROBLEM

Clearly, a fuel use problem exists with automotive power thermodynamics. And the resulting fuel misuse takes place with every vehicle there is, from the smallest minibikes to the largest diesel-electric trains, including all other powered vehicles in existence. The net effect of the vehicle fuel use problem is a demand for fuel energy to drive all vehicles, that is much higher than is required to actually drive the vehicles according to the mechanical laws of energy. This higher than empirically required vehicle fuel use is a 0% efficient, needless combustion of the highest volume of fuel and a total waste.

The fuel flow graphs, Graphs A, B and C illustrate that the highest fuel use rate takes place at the moment that engine power is applied to start vehicle acceleration. This fuel use takes place before the vehicle has started to move. The fuel use at the start of acceleration takes place when engine power is being converted into low grade heat to generate the drive force of acceleration needed to move the vehicle. This fuel use for the generation of a force for acceleration is totally in non-compliance with the mechanical law of motion W=FXD, Work equals Force times Distance. Before the vehicle moves, the "Distance," D, number of the calculation is 0, and therefore the calculated work done is also 0. Zero work done requires no energy according to the first law of thermodynamics, yet every vehicle on earth uses fuel at the highest rate, at the lowest efficiency, at this start of every vehicle acceleration. This is, by definition, fuel use at 0

% efficiency, at the highest volume of automobile fuel use, during every acceleration of every vehicle. We could not use fuel more wastefully.

The efficiency graphs, Graph D, E and F confirm this automotive misuse of fuel. At the start of vehicle acceleration, Graph D shows 0 % thermodynamic efficiency, as it should when all engine power is being converted to low grade heat just to generate the force needed to move the vehicle. When engine power is being converted to heat, it is, by definition, energy use at 0 % thermodynamic efficiency. Also, this 0 % efficiency fuel use takes place when engine power is near or at its highest output, at the beginning of vehicle acceleration and fuel flow is the highest. This means that all the fuel that is being used to operate the engine plus all of the generated engine power is being used at 0 % efficiency. None of the engine output power is being used for vehicle motion since vehicle motion has not yet started.

This application of force will start vehicle motion which requires mechanical energy. The engine output power provides this mechanical energy for vehicle motion and as such is the proper use of power. The portion of engine output energy that is properly being converted to vehicle motion, is no longer being downgraded to low grade heat to create drive forces as at the beginning of acceleration. As vehicle speed increases, engine output energy is being used as mechanical energy for vehicle motion. This process of shifting engine output energy away from force-based heat to motion-based energy continues as vehicle speed increases. Then as vehicle speed stabilizes, low efficiency transmission gearing stops and all engine output energy is being used for vehicle motion as it should be and is in compliance with the mechanical energy laws.

The essence of the fuel misuse problem is the speed discrepancy of drive-line components throughout the acceleration phase of vehicle motion. At the beginning of vehicle acceleration, the vehicle is at stop, and all drive-line components, wheels, axles, transmission parts and clutch are stationary. However, the engine output, the power supply for moving the vehicle must be in motion in order to supply power. No stationary mechanical device is able to supply power. Then as the vehicle acceleration begins, the moving source of power, the engine output is engaged with the stationary drive-line components of the non-moving vehicle. This non-synchronous engagement of moving engine and stationary vehicle drive-line components lies at the heart of the automotive non-compliant fuel use problem. This discrepancy of component speed requires force generation at unusually high engine power levels, and at low fuel use efficiency. This component speed discrepancy is present throughout vehicle acceleration, until vehicle speed is high enough so that vehicle component speed is moving fast enough to match engine component speed. At this point all drive-line component speeds are synchronized. Engine output mechanical energy supply is being delivered to vehicle motion mechanical energy demand. This takes place at 85 % efficiency and is the proper use of motive power energy, and takes place only at steady state highway speeds or at 1 to 1 engine to driveline synchronous speeds.

Clearly, today's automotive fuel use has a problem. The beginning of acceleration from stop of all vehicles uses the highest volumes of energy, both hydro-carbon or electric, at 0, (zero!) percent efficiency to start the acceleration process. Then as the vehicle begins to move, the energy use efficiency increases from zero (!) then 1%, gradually through all efficiencies and reaches the high of 85% for steady speed highway operations. This efficiency curve, from zero percent fuel use rising to eighty-five percent efficiency applies to all vehicles accelerating from stop to highway speed. But the laws of energy state that no energy, none, is needed to apply a force, while vehicles today use maximum fuel at zero efficiency to apply this drive force. Vehicles today use maximum when they should be using none! Clearly there is a problem.

Obviously, the science and technology for the proper use of motive power through the entire speed range of vehicles does not exist. It is the purpose of the documents that follow to provide the science and technology that solves the problem so that vehicle fuel use equals or is in alignment with the actual energy needed to accelerate the mass of the vehicle. If this is accomplished, the benefits would be enormous. Imagine!

THE SCIENCE OF MOTION

INTRODUCTION

The science phase of this paper is an essential requirement if automotive fuel use compliance with the laws of energy is to be accomplished. The science required to guide automotive drive-line engineering technology does not exist. Automotive manufacturers have no science to which they can refer so that the appropriate engineering can design automotive drive-line systems that comply with the laws of energy. Without the science available, the automotive industry has been able to engineer motive systems for all vehicles, large or small, that operate conveniently at transportation, but at the cost of very high, non-compliant fuel consumption.

The previous problem phase outlined and clarified the essence of automotive thermodynamics in mechanical terms, and explains the reasons for the inordinate fuel use.

The understanding of this problem provides the ground work for the research to discover, clarify and state, the new physics needed for automotive energy compliant drive-line design. The purpose of this science phase is to present the physics that does not yet exist, so that automotive design engineers can start to move toward automotive fuel use compliance.

The research for the new physics is done in the analysis section of this science phase. The beginning of the analysis is our present understanding of the science of motion. This will, necessarily, refer to Newton's laws of motion. Then to further develop the science, the characteristics of motion that we do not understand will have to be defined. A source of energy in the science of motion, will have to be defined in empirical terms. The characteristics of the moving body will have to be made clear. The dynamics of the relationship between a source of energy and a moving body that is being driven by that source of energy, will have to be defined. Every aspect of the essence of the automotive fuel use problem will have to be addressed in the analysis, if the new physics is to be relevant and useful. The findings will have to confirm the use of energy for driving a forced, variable speed body. The analysis will explore the relationship of the new findings with existing physics, namely the laws of motion. The relationship between the existing physics and this new research provides the ground work for the new laws of motion that follow.

After this, the statement of the Fourth Law of Motion follows. The specifications of the law are explored. Its relationship with reality is defined so that its relevance to solving the compliance fuel use problem is merited. Its **qualitative** characteristic is recognized and this provides the framework for the Fifth Law of Motion. The statement of the Fifth Law of Motion provides the **quantitative** definition of the Fourth Law. A previously existing equation, the mechanical law of energy, provides the means for quantifying the Fourth Law, and requires a fuller definition of the equation to qualify it as the Fifth Law of Motion.

The document that follows discusses the analysis needed for the development of these new laws of motion. Later, the new physics will conclude the science section of the book.

Scientific Analysis

Newton's laws of motion provide the framework for our understanding of the science of motion. Newton's first law is:

"A body at rest or in motion, remains in that state, unless acted upon by a force."

It defines the inertia of a body. This first law of motion is the very definition of inertia. This law applies to all matter that exists, including every vehicle. All vehicles will remain motionless if they are not acted upon by a drive force. This first law provides the qualitative scientific foundation for the science of motion.

Newton's second law of motion defines "force." It states:

"The acceleration of a body varies directly with the force, and inversely with the mass of the body." ie "F = M x A."

The force, defined by this law, has an interactive relationship with both the "acceleration" and the "mass" of the body. The application of a force on a body initiates the acceleration of the body. As long as the drive force is being applied, the motion of the body will be changing speed, either speeding up or slowing down. The force initiates a steady increase in speed of the body, which is more than just the uniform motion of the body, where no force is required, according to the first law.

The acceleration of the body can be defined by observation and therefore this aspect of force can be known in empirical terms. But the other aspect of the forced body that effects acceleration, is the mass of the body. According to this second law, the more massive the body, or "heavier," if you will, the slower the acceleration of the body for any given applied force. And the less "full of mass" or "dense" the body, the faster the acceleration for any given applied force. The resulting acceleration of the applied force, varies inversely with the mass of the body, as the law states. The mass, being a physical property, can be verified by observation, and therefore, the specification of force can be defined in empirical terms.

Given that both mass and acceleration can be defined in empirical terms, the force of the second law can be defined. The force is defined by the equation, "Force equals mass multiplied by acceleration" or, in abbreviated form as F= M x A. This equation, the quantitative definition of the Second Law, together with the qualitative definition of the First Law of Motion provide the foundation for the science of motion. But Newton understood mass well enough to know that there was an interaction between the mass and the applied force, hence the Third Law of Motion.

Newton's next, Third Law of Motion is:

"Every action has an equal and opposite reaction."

This is Newton's Third Law of Motion. When a force is applied to a body with mass, the applied force is met with resistance to motion, due to the inertial mass of the body. This resistance to motion opposes the applied force in direction and magnitude. Therefore, the law states "opposite and equal" in the description of the interaction between the applied force and the reaction force returning from the mass. And this action-reaction relationship takes place with every applied force onto a body with mass.

These three laws from Newton provide the fundamental physics for all motion. More specifically, and according to our purposes, these laws apply to all vehicle motion. The motion of every vehicle, follows the First Law. A vehicle moves only when acted upon by a force. The acceleration of the vehicle follows the Second Law and moves according to the applied motive force on the vehicle, and its resistance to motion, due to its mass. The reaction forces to this vehicle motion are harnessed by the vehicle drive train. The vehicle drive train provides the resistance of the vehicle reaction force, as required by the third law, but also provides the energy required to move this force through a displacement as is required by the mechanical laws of motion. Without the energy provided by the drive line, continuous vehicle motion would not be possible. It is from this required application of energy to a forced, moving body, where Newton's laws of motion do not cover the physics. This is the point of departure for further research in an effort to find the laws of motion that describe the physics of motion and the introduction of energy in the application of force to a movable body.

ENERGY COMPATIBLE LAWS OF MOTION

To be able to determine the relationship between a source of energy and a moving body, the motion of both needs to be understood. The characteristics of motion of the two primary sources of power for vehicles, electric motor and fossil-fueled internal combustion engine, need to be discussed. Then the motion of a forced, moveable body, such as any vehicle, needs to be understood. The comparison of these motions follows. Insights from the understanding of energy non-compliance from "The Problem" will be used to further the understanding of the compliant relationship between energy and motion. After this is understood, the next law of motion, the Fourth Law, is stated.

The essence of a source of mechanical energy is motion. The equation "work equals force multiplied by distance," or $W = F \times D$, or the mechanical law of energy proves this understanding. Work represents energy and whenever the distance, D, is zero, the resulting work is zero and the required energy is zero, regardless of the intensity of the applied force. Due to this law of energy, displacement or motion must be available from a prime mover if it is to provide energy to a moving body or vehicle. An engine needs to be running, with a moving output shaft before it can supply the energy to move the vehicle. An electric motor must be able to move or operate if it is going to drive anything.

Prime Mover Motion

The characteristics of the motions from sources of energy, engine and motor, are different. The output shaft, or crankshaft of the engine must be at least at idle speed for energy to be available from the engine. A "stopped" engine is not able to supply energy to a load. Engine speed must be increased to a level able to supply enough power when engaged with the load of moving the vehicle not to stall. This application of drive force to the vehicle is always, with all vehicles, a moving shaft engaged with the non-moving vehicle driveline, and the resulting non-compliant energy use. Electric motors must also supply motion in order to provide energy for vehicle motion. Electric motors are variable in speed and are able to supply torque force at zero speed. But energy use of a "stalled" but torqued motor is high, at high voltage and amp draw that needs to be controlled. The zero-speed, electric motor has no back electro-motive force, back EMF, and amp draw at zero speed full torque is at max. Even though the required motion for power delivery from an engine is very different than the motion of the electric motor, they both use fuel or energy flow according to Graph A, and in non-compliance with energy laws. Both electric motor and engine energy use, upon the engagement of vehicle drive force, use energy to apply a drive force on a load. Both electric and engine driven vehicles use energy to apply force to the vehicle load in non-compliance with W = F x D. The non-compliant use of fuel or energy is the result of force application to a non-moving vehicle, directly because engine power is at speed and electric power is at stall. This is all vehicles. The science regulating a powered source of energy applying a drive force does not exist and needs to be developed to be able to move into compliant energy fuel use. The application of a drive force onto a non-moving load requires no energy and laws governing this do not exist.

Vehicle Motion

Vehicle speed change, faster or slower vehicle motion, is also a characteristic of prime mover motion. Both engine and motor output shaft speeds operate in proportion to vehicle speed. The engine output shaft speed engages with the vehicle drive-line through the clutch and transmission. When the clutch is fully engaged the engine output shaft speed is driving the vehicle through the lowest gear ratio of the transmission. This is a fixed gear ratio, say first gear, and vehicle wheel speed, ergo vehicle speed, is proportional to engine output speed. This fixed proportionality requires engine speed, rpm, to increase for the wheel speed, ergo vehicle speed to increase. As the engine speed rises, and reaches a higher operating rpm, the transmission is shifted into its next gear ratio, where the speed change process starts all over again, for both engine and vehicle. This speed change process continues through all of the transmission "gears" until both the desired vehicle speed and the efficient engine speed have been reached. Slowing the vehicle, other than braking, requires the slowing of engine speed to slow vehicle speed. The slowing of the vehicle is the opposite process of increasing vehicle speed, but both are fixed proportional speed changes through the gear ratios of the transmission.

Motor speed must also change for electric vehicle speed to change. The motor is directly coupled to the vehicle drive-line, without a clutch, due to its capacity to torque at zero rpm, and at zero vehicle speed. As force is applied and vehicle speed increases, electric drive motor speed, also increases. Electric motor speed must increase as vehicle speed increases, due to the fixed proportionality ratio of the motor and vehicle wheels. The motor drives the vehicle without a range of ratios or a clutch as is required by the I.C. engine driven vehicle. But to control the vehicle speed, the motor speed must be controlled by the vehicle operator through a frequency modulator or a pulse generator depending on design. These are able to operate the motor through the entire range of motor speed, from zero to top design speed. This speed change of the motor, through the fixed drive-line design ratio is capable of moving the vehicle from stop to design highway speed, the desired objective. Slowing the vehicle reverses the acceleration process. The controller slows the motor speed through the fixed ratio drive-line that then slows wheel speed and vehicle speed. This process often regenerates battery charge. Overall, the state of the art for automotive transportation is the proportional relationship between prime mover output shaft speed and vehicle speed.

The multi-purpose use of vehicle prime mover shaft motion is noteworthy. Prime mover motion is essential for the delivery of energy to the moving vehicle. This is dictated by the laws of energy and is beyond debate. But, prime mover change of speed is also needed for today's vehicle motion change of speed. Prime mover motion for the use of speed change may be addressed, if the new science that follows, frees it from slavish service to vehicle speed change. The requirement of an internal combustion engine to be variable in operating speed limits its thermodynamic efficiency. Freeing prime movers from speed change duties would allow much needed optimization in design and improvement in thermodynamic efficiency.

Motion Comparison, Prime Mover and Vehicle

In order to further develop the motion relation ship between prime mover and load, we now need to understand load motion, or vehicle motion. Vehicle motion and vehicle drive-line component speed are directly related. At zero vehicle speed, wheel speed, and all drive-line components are also at zero speed. This is the physical condition of vehicle components at the application of force for the beginning of every vehicle acceleration. Then as vehicle speed rises above zero, the motion requires that drive-line components and wheel speed begin to move at a rising rate. The faster the vehicle moves, the faster the wheel speed and drive-line components need to move. Drive-line speed and vehicle motion are directly proportional, based on component gear ratios and wheel speed specifications. Then when vehicle motion reaches steady state highway speed, drive-line components and wheel speeds have also reached their steady state operating speed. And as long as the vehicle moves, drive-line components move accordingly. This proportional relationship is undebatable, and translates vehicle motion into interaction with prime mover motion at the vehicle transmission.

The interaction of the prime mover motion and the vehicle drive-line motion determines energy use efficiency. Vehicle drive-line motion varies from zero speed to highway speed at all times. This means that drive-line component speed varies from zero to highway speed at the transmission at all times. This is its only range of motion. On the other hand, the internal combustion engine must always be operating and in motion to be able to supply energy for vehicle motion. The electric motor drive must also be in motion to supply power, but has been engineered to be able to supply torque at all speeds from zero to highway speed in order to simplify motion interaction with vehicle driveline components. Both the engine and the motor have been enslaved to be variable in speed to match vehicle speed range from zero to highway in order to simplify power to load interaction at the transmission. This simplification at the transmission is the source or all non-compliant automotive energy use as outlined in the problem section earlier. Further development of the prime mover motion and vehicle motion interaction has never taken place, due to the lack of physics and the laws of motion governing such engineering.

The essence of the breakdown of the laws of motion at the transmission is the requirement of prime mover motion, even at times of zero vehicle speed. The required mechanical motion of the engine at the engagement of a zero-speed vehicle drive-line is the source of all automotive energy use non-compliance as highlighted in the problem section earlier. And, non-compliant energy use takes place with electric drive at zero speed as noted above. The engagement of prime mover motion with non-motion driveline at the transmission for the application of the accelerating drive force is in non-compliance with the mechanical law of energy. This motion to non-motion interaction is in complete compliance with the third law of motion for the application of a force onto a load. But it does not provide the knowledge required for energy compliant motion, a needed favorable characteristic for solving today's automotive transportation energy use problem.

Nugget

The next law of motion, The Fourth Law, capable of regulating energy, needs to be developed. Newton's Third Law sets the stage for motion to take place. It recognizes that for the motion of a body to take place, a force has to be applied to the body. In the application of the force to the body with mass and inertia, a reaction force is generated by the body equal to and opposite in direction to the applied force. This reaction force provides resistance to the original applied force. It is the mismanagement of these reaction forces in today's automotive drive-lines that is the source of the non-compliant use of energy, that the Fourth Law must correct.

The mismanagement of the reaction forces of the Third Law of Motion was identified in the problem section earlier in the paper, and also, above. The essence of the misapplication of forces in all automotive drive-lines is that the source of energy or power is moving when engaged with the non-moving vehicle drive line. This engagement of the "moving" with the "non-moving" uses energy when, the application of a force, theoretically, does not require energy in order to be applied. The essence of the problem is that the reaction forces of the non-moving, stationary vehicle drive-line components do not have a stationary foundation or bracing frame-work to harness or provide resistance to these forces. If a stationary foundation was available for the stationary Third Law reaction forces, the non-moving reaction forces would act upon non-moving bracing, and the force could then be applied without the use of energy from prime movers, and in compliance with the laws. The Fourth

Law of Motion must make this provision available in order to correct the existing non-compliant fuel use and rise above present-day automotive technology state of the art.

Secondly, the Fourth Law has to make provision for the moving body reaction forces. The motion of the moving body will be defined in relation to the motion of the source of energy required for the motion to take place. When the motion of the moving load is equal to the motion of the prime mover, the load is considered a moving body. At this point, the transition from stationary is over, and the motion of the body has reached steady state. And as a moving body, reaction forces, equal to and opposite to the applied forces on the moving body are generated as with any applied force, stationary or moving. This mode of operation in the automotive technology state of the art is in compliance with mechanical laws of energy. The moving prime mover matches the motion of vehicle components and in so doing, supplies the energy required for motion of the vehicle. The prime mover carries the reaction forces of the forced and moving body. Here the Fourth Law of Motion does not need to correct today's automotive state of the art of a moving vehicle. But it must manage the reaction forces of the moving body to be in compliance with the mechanical law of energy above and below this 1 to 1 engine to vehicle relationship.

Thirdly, the Fourth Law has to make provision for the proper management of the reaction forces generated by the body during acceleration or the transition from zero to steady state speed. The applied force to the body to start acceleration, generates a reaction force that needs to be managed during the body's speed transition from zero to steady state motion. Today's automotive state of the art applies the accelerating force by the engagement of the moving prime mover with stationary vehicle transmission components, thereby using energy for the force application in non-compliance with laws. As noted, this is the incorrect application of fuel dependent force at the beginning of vehicle acceleration from zero. Then as vehicle speed increases from zero toward steady state full motion speed, the first misuse of fuel for the application of force for acceleration begins to be used correctly, for the actual motion of the vehicle. This is shifting fuel use away from mis-use, at the beginning of acceleration to proper use in the driving of the vehicle, during acceleration, and at 1 to 1 engine to vehicle ratio. This, today's, automotive technology state of the art, moves from full stationary non-compliant fuel use to full motion compliant fuel use during the transition from zero to full speed. The Fourth Law of Motion must correct this fuel misuse and correctly manage reaction forces during vehicle speed transition from zero, or 1 to 0 ratio, to full speed, or 1 to 1 ratio, in order to achieve compliance with mechanical energy laws.

This management of the reaction forces by the Fourth Law of Motion during the transition from zero to full speed is critical. At zero vehicle speed, the correct placement and control of the zero-speed reaction force is on a fixed or zero speed framework or support. This allows the acceleration force to be applied to the non-moving vehicle without the use of energy as is required by the mechanical law of energy. Then as the vehicle begins to move, and energy is required to do the work of forced vehicle motion, a portion of the force resting on the fixed support must be transferred onto the moving shaft of the prime mover to supply the needed energy. This means that a portion of the reaction forces will be carried by the fixed support, and a portion of the reaction forces will be carried by the moving support of the prime mover, thereby providing energy for vehicle motion. The size of the portion remaining on the fixed support will be determined by the needed reaction forces transferred to the engine for the proper amount of energy to drive the vehicle according to applied force and vehicle speed. Then as vehicle speed continues to increase, more of the reaction force must be transferred from fixed support to engine's moving support to supply the increasing energy requirements. This continues until vehicle speed reaches full, as defined above, and all reaction forces are removed from the fixed support and are fully transferred onto the moving prime mover out-put shaft, when vehicle speed has reached full and acceleration is finished. This transition of reaction forces from fixed to moving must be properly managed by the Fourth Law of Motion for the acceleration and drive of a vehicle to be in compliance with the mechanical law of energy and motion.

Summary

Ultimately the Fourth Law of Motion must make two contributions over and above Newton's first three laws of motion. Firstly, the Fourth Law must manage the reaction forces defined by the Third Law, so that all applications of the reactions comply with the proper use of energy as defined by the mechanical law of energy. The proper management of these reaction forces requires that the Fourth Law corrects energy misuse at vehicle zero speed, and also corrects the energy misuse during vehicle speed transition from zero to full speed. These required corrections rise above today's automotive technology state of the art energy misuse for vehicle motion. Ultimately, all reaction forces, in the forced motion of a vehicle have to be properly managed for compliant energy use, regardless of vehicle speed. Secondly, as an integrated benefit of the corrected and proper management of the reaction forces, the Fourth Law must define the correct use of energy in the forced motion of a body, such as a vehicle. This defined use of energy does not exist in the Newton's first three laws, and is a necessary requirement in the laws of motion, and is the Fourth Law's reason for existence.

PART TWO, THE FOURTH LAW OF MOTION

This is the introduction to the Fourth Law of Motion. Here, Newton's Third Law of Motion serves as a natural context for the Fourth Law. It provides the connection between Newton's first three laws of motion to the new Fourth Law. It also provides the opening of the discussion for the Fourth Law, since it is the beginning of the process of motion by introducing both concepts, "action" and "reaction." These then provide the setting needed by the Fourth Law to introduce the concept of energy required by any action or force which is driving a mass through a distance.

Newton's Third Law of Motion states that: "Every action has an equal and opposite reaction." This is the beginning of the motion process. The action, referred to by the Third Law, is the applied force needed for the motion of a body to take place. The state of the body would remain unchanged, according to the First Law, without the action of the Third Law applied to it. The action, the applied force, changes the state of the body, according to the magnitude of the applied force and inversely according to its mass. This is Newton's Second Law of Motion. The mass of the body provides the resistance to motion that generates the reaction forces when the forced action of the Third Law is applied to the body. This then changes the state of the body, and generates the equal and opposite reactions of the Third Law. However, the moment the state of the body changes due to the action, the applied force, the force has been displaced through a distance and energy is required for the change of state of the body to take place. At this moment, the physics of Newton's laws of motion is inadequate for the supply of energy needed for the body's change of state. It is the purpose of the Fourth Law of Motion to provide the natural advancement from Newton's first three laws of motion, to describe the physics of energy needed for motion to take place in compliance with the laws of energy.

THE FOURTH LAW OF MOTION

The statement of the Fourth Law of Motion is as follows:

Equal and opposite reactions, act upon fixed bodies, moving bodies, or proportionally, both fixed and moving bodies.

As stated earlier, The Fourth Law must be capable of providing the necessary physics in order to be relevant and useful. It's starting point has to be the Third Law of Motion. It has to provide bracing and support for all reaction forces generated throughout the vehicle's change of state, from zero to full speed, weather in steady state or in transition. And, integrated with the process of the proper management of the reaction forces, it must supply the energy needed for the speed change of the vehicle, but in so doing, it must supply the correct energy needed for the change of speed. Due to the integrated nature of the Fourth Law, if any one aspect of its provisions is off, none of the others would be in compliance. They all work in compliance or none work in compliance.

EQUAL AND OPPOSITE REACTIONS

The Fourth Law of Motion begins with, "Equal and opposite reactions." It is the natural beginning for the Fourth, since the equal and opposite reactions are the outcome, the end result of the applied action of the Third Law. The Third Law can go no further in the concept of motion than to apply the force and thereby generate equal and opposite reactions to that force. The resulting equal and opposite reactions of the Third, provide the setting, for the Fourth Law to begin to apply its provisions. Upon the availability of the reactions, the Fourth Law must apply its provisions to any set of circumstances between the source of energy, the prime mover shaft and the moving body, the load, regardless of their speed relationship or applied force. Any and every circumstance in the defined relationship between the source of energy and the load, has to be integral with the Fourth Law so that its provisions are relevant and in total compliance with the laws of energy. This elevates the provisions of the Fourth into requirement statis for compliance.

The relativity of the motion of the source of energy and the motion of the body needs to be defined so that the action/reaction relationship between the two can be correctly managed. A non-moving structure of the source of energy, the driveline, or the transmission case usually mounted to the frame of the vehicle will serve as the fixed frame of reference or the fixed body. This frame of reference will define "fixed" when the speed of the load is compared to it. So, when the driveline elements of the vehicle, are compared to the fixed frame of reference, and they are both stationary, at the same speed of zero, the load is considered to be at zero speed. Any reaction forces resulting from the process of motion from the driveline elements of the vehicle at zero speed can be transferred to and supported by this fixed frame of reference.

Similarly, the "moving body" needs to be defined. The moving element of the source of energy from the engine, the crankshaft, say, will be considered the moving frame of reference or the "moving body." Whenever the vehicle driveline elements are in motion, such that their speed has stabilized, and the relative motion of the load and the engine are similar, both in motion, the equal and opposite reaction forces of the driven load could be transferred to the engine for the proper action/ reaction relationship. In this setting, the moving body, the engine output shaft would bear the reaction forces of the steady speed vehicle load, so that the correct energy needed for driving the load would be drawn from and supplied by the engine. This could be the upper end of the vehicle speed scale, where no further acceleration is required for the motion of the vehicle.

The third frame of reference for the application of equal and opposite reactions needs to be defined. The first frame of reference defines steady state at zero speed. The second frame of reference defines steady state at designated steady moving speed. The third frame of reference, the transition of speed from zero to full speed, needs to be defined. This frame of reference is functioning, during the acceleration of the load. At the beginning of this process, all reaction forces are borne by the fixed body. At the end of acceleration, all reaction forces are borne by the moving body. During the accelerating speed transition process, the reaction forces need to shift away from the fixed body, and onto the moving body. The reaction forces during this phase are partly borne by both the fixed body and the moving body frames of reference. The rate of shift will depend on the energy needed for driving the vehicle, and the intensity of the applied driving force. The reaction forces will be split between fixed and moving. The proportionality of the shift will depend on energy required by the motion of the load versus reactions from applied forces returning from the load. This third reference system completes the standards needed for the proper management of the action/reaction relationship for the motion of a body. These are the conditions provided by the Fourth Law so that compliant fuel use for vehicle transportation is possible. Compliant fuel use under the control of the Fourth Law must be both corrected and aligned with energy requirements at all vehicle speeds.

Fixed Bodies

The Fourth Law defines the action/reaction relationship between the load and the source of energy. At the beginning of the motion process initiated by the Third Law, the load, the mass of the vehicle, is stationary when the action, the force, is applied and the reaction forces are generated. The Fourth Law requires that these reaction forces are borne by a fixed body of equal motion as the "fixed" load. This requirement prevails, even though any source of energy must be moving in order to supply energy, even though there is a discrepancy of motion between the engine and zero vehicle speed. This action/reaction relationship must exist at the beginning of forced acceleration so that the source of energy, the engine, does not need to provide energy during the application of force. This application of force, that does not require energy is in compliance with thermodynamics. So, the provision of the fixed body that bears the reaction forces from the non-moving, fixed vehicle is the requirement of the Fourth Law even though the engine is operating and in motion.

This is a fundamental correction provided by the Fourth Law over and above today's automotive state of the art. The Fourth law requires the provision of a non-moving foundation to bear the reaction force of the non-moving load, the vehicle, so that the correct amount of energy is used for the application of the force. No energy is needed to apply a force onto a load. This requirement is over and above today's automotive standard where the reaction force of the action applied to accelerate the vehicle is borne by the operating, engine moving shaft and is, thereby, using energy for the application of the force. This problem is clearly outlined in the problem section of this paper. The use of energy for the application of a drive force is in non-compliance the laws of energy. The Fourth Law corrects this violation and upgrades standards into compliance with mechanical laws of energy.

Theoretically, moving bodies do not need a supply of energy to maintain motion. This is Newton's First Law of Motion. But, specifically, when trying to understand vehicle motion, its obvious that steady state vehicle motion can only be maintained with a supply of energy, due to vehicle rolling resistance. A vehicle, given steady state motion without a sustaining supply of energy will eventually roll to a stop. This is due to the many resistances to motion every vehicle possesses. Energy for vehicle motion requires the moving bodies frame of reference provided by the Fourth Law. Indeed, the Fourth Law of Motion, and its provision of the moving body source of energy, are more easily understood when related to maintaining vehicle motion.

But, sustained, steady speed vehicle motion must overcome many resistances to motion. The aerodynamic drag of the vehicle's body shape constantly tends to slow vehicle motion. This drag varies directly with vehicle speed. The higher the vehicle's steady speed the higher the aerodynamic drag, the more energy is required to maintain steady state speed conditions. All vehicles, regardless of their streamlined design, suffer this resistance to motion. Also, the tires that vehicles roll on, have a rolling resistance. The standard air pressure in tires allows them to flex as they support vehicle weight. This flexing generates heat and requires energy to overcome. Also, the internal losses of a vehicle's driveline system increase rolling resistance. The internal parts of the transmission and differential wheel drive operate in a constant lubricating oil bath. The viscosity of the oil through which the gears operate, raises the drag on the driveline, generates heat, and requires energy from the engine to overcome. With other sources of rolling resistance, vehicle motion can only be sustained with a supply of drive energy from the engine, the moving body frame of reference, provided by the Fourth Law.

The Fourth Law requires that the reaction forces generated by the vehicle driveline elements, when at steady top speed, be borne by the engine crankshaft, the moving frame of reference. This defines the proper management of the reaction forces, at the higher levels of the speed range of the vehicle. This handling of the reaction forces insures that the energy required for the drive of the vehicle is in compliance with the laws of energy. This occurs at the moment that vehicle acceleration has come to an end, steady speed is reached, and the applied forces that have initiated acceleration have been lowered to just equal steady state rolling resistance. This action results in the full application of reaction forces onto the source of energy, the engine, the moving frame of reference. But depending on ratios and defined vehicle speed, proportional sharing of reaction forces onto the fixed body frame of reference is still possible.

PROPORTIONALLY, BOTH FIXED AND MOVING BODIES

Fixed bodies and moving bodies define the action/reaction relationship with frames of reference at the beginning and the end of vehicle acceleration. However, the management of the reaction forces during acceleration, during vehicle speed transition from zero to full, is the third requirement of the Fourth Law. Both "fixed" bodies and "moving" bodies, each refer to a single frame of reference, fixed and moving. However, the acceleration or speed transition phase of vehicle motion, requires a proportional reaction force relationship with the two frames of reference at the same time. These two frames of reference are the fixed body and the moving body provided, required, by the Fourth Law.

The acceleration phase of a vehicle requires the management of the action\reaction relationship with two frames of reference, fixed, and moving. The moment the vehicle begins to move, the reaction forces begin to shift. The reaction force that is borne by the fixed frame at zero vehicle speed, begins to move onto the moving frame of reference as motion starts and energy is needed. At the beginning of acceleration most of the reaction forces are still borne by the fixed body and only a small amount of reaction force is on the moving body, the engine. This small amount of reaction force borne by the engine is equal to the energy needed to move the vehicle at slow speed, all in compliance with energy laws. As acceleration continues, more of the reaction forces move away from the fixed body and onto the moving body frame of reference, the engine, as more energy is needed to drive and accelerate the vehicle. As acceleration ends, and steady full speed is reached, the shifting of the reaction forces is also at steady state fixed body/moving body proportionality. When vehicle speed has reached steady state, and depending on system ratios, the load's reaction forces could be borne by just the moving body, or by the proportional interplay of both fixed and moving bodies.

The proportionality of the placement of the reaction forces, onto fixed and moving bodies, during the acceleration of the vehicle, is a requirement of the Fourth Law. This requirement during vehicle acceleration is essential in order to ensure compliance with the laws of energy. As the vehicle begins to be moved by fully intense drive forces, the portion of the reaction forces to be transferred on to the moving, energy supplying engine is very small, for the energy needed, even though the reaction forces are still fully intense. The large portion of the reaction forces still unused to extract energy from the moving frame of reference engine, must still be borne by the fixed frame support system. Then as vehicle speed increases, a higher portion of the available reaction forces will be borne by the moving frame engine, to fill the added energy compliance needs of the higher speed. As a result, a smaller amount of the reaction forces will be available for transfer to the fixed frame of reference. Then, at steady full speed, all of the reaction forces are borne by the moving, energy supplying engine, and no further portion of the reaction forces are available for fixed framed restraint. At this point, as throughout the process, energy extraction from the engine equals the energy needed to accelerate and drive the mass of the vehicle. This energy compliance is only possible because of the proportionality requirement of the Fourth Law.

When vehicle speed increases so that more energy is required than the 1 to 1 load reaction force/moving body relationship can provide, proportional interaction with the fixed body begins again. At ratios above 1 to 1, the fixed body/moving body interaction becomes reactionary. The regular drive forces needed for the higher vehicle speeds are reacted back from the load to the fixed and moving bodies. But in order for the higher energy to be extracted from the moving body engine, the fixed body has to provide the reaction force equal to the higher load-force applied to the prime mover moving body. The load's regular reaction forces, communicated back to the moving body for higher energy extraction, requires equal higher reaction forces be borne by the fixed body, in order for the moving body to provide the added energy. The fixed body and the moving body provide each other's reaction forces for the energy needed at higher vehicle speeds. And so the proportional fixed body/moving body relationship continues for ratios above 1 to 1.

Proportionality of reaction force placement onto the fixed and moving bodies varies directly with vehicle speed, according the Fourth Law. This is the second fundamental correction provided by the Fourth Law, over and above today's automotive state of the art. The Fourth Law requirement that the proportional bracing and support of the reaction forces onto both fixed and moving bodies, simultaneously, prevails, even though the speed difference between them is at maximum. The demands of this energy compliant specification of the Fourth Law has been too difficult for automotive design efforts to accomplish. As a result, both reaction forces, force based and energy based are borne by today's engines, using fuel to apply force, totally contrary to the laws of energy and thermodynamics. The Fourth Law requirement that the two characteristic reaction forces, force and energy, i.e. force at motion, act proportionally to both frames of reference, fixed and moving, must be followed completely, if energy compliance is to be achieved, regardless of how difficult it may seem. The Fourth Law makes no allowances for ease of achievement, in its requirements, and expects the physical world to comply, in order to align fuel use with actual automotive transportation energy needs.

A Summary of The Fourth Law of Motion

The purpose of the Fourth Law is to provide the physics needed, so that the proper use of energy for all transportation can be accomplished. Logically, it must continue from the setting provided by Newton's first three laws of motion. The reactions provided by the Third Law of motion is the starting point for the Fourth Law. The Fourth Law then manages the placement of the reaction forces according to all vehicle speeds in order to achieve compliance to the laws of energy. This proper placement of the reaction forces corrects two fundamental errors of fuel misuse, over and above today's automotive state of the art. The required placement of the reaction forces by the Fourth Law is exact and this precision guarantees compliant fuel use at all vehicle speeds. In so doing, the Fourth Law fulfils its purpose and provides access to the correct energy use for transportation and eliminates huge amounts of wasted fuel use. Carbon foot print, due to this correction of automotive fuel use, may be reduced to sustainable balance as a result.

The essence of the breakthrough of the Fourth Law is that it defines two frames of reference, "fixed" and "moving," bodies used to communicate the reaction forces from the driven load back to the drive train and source of power. The fixed body refers to the non-moving drivetrain element. The moving body refers to the prime mover source of power output. The interplay of these two elements allows the compliance to energy laws according to the ratios of the input and output shafts, at the transmission. This also requires the management of two energy transmission systems, each related to either fixed or moving bodies of reference. At the lowest ratio, 1 to 0, the reaction forces are borne by the fixed transmission system and body, according to the law. As ratios increase, reaction forces from the load are translated to both fixed and moving bodies by both energy transmission systems. The interaction of the fixed and moving bodies, each with its own force management transmission system, continues throughout the range of input/output ratios, totally incrementally, all in accordance with energy laws. Every reaction force from a driven load, is communicated to the driveline so that alignment of energy use and energy need for the driven mass is guaranteed, according the Fourth Law of Motion.

Segue into Technology

The effort to bring the Fourth Law into physical existence requires systemic clarification. Relevant operations of physical technology, and its relationship with theory, needs to be understood. The range of ratios of the technology must be from the lowest ratio, 1 to 0, to the highest ratio, infinity, or, 0 to 1, that is possible between an input shaft and an output shaft of the load driving system. The only parameters available for defining the limits of this range of ratios, are the two frames of reference defined by the Fourth Law, fixed and moving bodies. The lowest ratio communicates the load's reaction forces back to the fixed body for compliance. The infinity, highest, 1 to 0, ratio communicates the load's freewheeling reaction forces back to the same fixed body frame of reference, as with the lowest ratio. This means that the limiting parameters for the whole range of ratios for technology operations is the same fixed frame of reference as defined by the Fourth Law. This is not the usual defining parameters for a range of behavior, that usually refers to two or both possible, available references, one for each end of the range. But the requirement for compliance to W=F x D demands that the range of ratios is defined by the fixed frame of reference at both ends of the range of ratios.

But technology operations must include the moving frame of reference for energy to be applied and available, when the load's reaction forces are communicated back to the driveline, and while the driven load is moving and needs energy to move. This can only fully take place when no proportional interaction between fixed and moving bodies needs to take place during the range of ratios and all reaction forces are borne by the prime mover moving body. This must be at the 1 to 1 ratio, exactly at midrange of the ratios. At all other ratios, there is the proportional interaction between the fixed and moving bodies, shared or reactionary, in order to maintain compliance with the controlling theoretical mechanical law of energy.

Technology operations working in compliance with the Fourth Law, must have the full range of ratios, from the lowest to the highest. The defining parameter, (singular) for both ends of this range of ratios is the fixed body frame of reference defined by the Fourth Law. The technology must have access to the entire range of ratios, by shifting the load's reaction forces proportionally between both the fixed and moving bodies. At midrange, the fixed body interaction is not needed, and reaction forces are fully borne by the moving body prime mover, at the ratio of 1 to 1. As ratios increase, the Fourth Law fixed body becomes proportionally engaged with the moving body, extracting energy as needed for the increasing work load, until the highest ratio, 0 to 1, or until the infinity ratio is reached. At infinity ratio, the fixed body blocks all moving body motion and the load freewheels, resulting in the 0 to 1, or infinity ratio. The real-world technology operations must work according to this range of ratios and the limiting parameters. These physical operations will then be controlled by the theoretical energy equation, W = F x D. Due to the theoretical context and control of this equation, over all physical operations in the range of ratios, energy compliance is guaranteed. We can now build the technology that is the manifestation of the Fourth Law in the real world.

The Fourth Law is qualitative in nature. To gain a fuller understanding of it, the Fifth Law of motion with a quantitative description of the Fourth Law, has to be defined. This follows Newton's lead in his writing of the First and Second Laws of motion. Newton's First Law is qualitative in nature, and his Second Law of motion is quantitative in nature. The following description of the quantitative Fifth Law finishes the science section of this paper.

THE FIFTH LAW OF MOTION

This is the introduction to the Fifth Law of motion. This definition is needed to provide the quantitative understanding of the Fourth Law, that we now understand qualitatively from the previous section. The Fourth Law provides the knowledge about the appropriate use of fuel, when used for transportation. The quantitative Fifth Law of motion defines the amount of fuel energy needed for the acceleration and drive of a vehicle from zero speed to steady state high speed, or higher, whatever that may be.

The energy required to accelerate the mass of a vehicle varies directly with the applied drive force causing the acceleration and the distance through which the force is applied. Essentially, the energy required to accelerate the vehicle is defined by the mechanical law of energy.

Therefore, the statement of the Fifth Law of motion is as follows:

Work equals Force times Distance, or W=F x D

This equation is found in every high-school physics text describing machines and mechanical energy. The added credential of "Fifth Law of Motion," to this equation, will provide the quantitative understanding of energy needed for the work done during the acceleration of the mass of the vehicle. The Fourth Law ensures that the energy equal to this equation is extracted from the source of energy, the prime mover, for the acceleration of the mass of the vehicle.

The elements of the equation need to be understood. The symbol W represents "work." F represents "force." D represents "distance." The numerical value of force can be in pounds. The numerical value of the distance can be in feet. The value of the work energy can be understood in foot-pounds. The work energy, according to the equation, is the product of the multiplication of the force and the distance. The work energy exists when a given drive force is applied to a mass, and the mass moves a certain distance while being forced. The product of the force and the distance, define the work energy applied to the mass. This means that, according to basic mathematics, when either the force or the distance, are a value of zero, the value of the work energy is also zero. When the distance traveled is positive, but no applied force was used for the motion to take place no energy was used. When the applied force to the mass is high and positive, but the mass has not moved, no energy is used in the application of the force. This is the meaning of the equation. It holds the Fourth Law to energy compliance protocols.

It is the understanding of this relationship, that judges the automotive fuel misuse standards in the harshest terms. The equation says that the application of a drive force on the mass of a vehicle at the beginning of acceleration, before any distance has been travelled, requires no energy! Distance is zero, and the product work energy is therefore zero. But today's auto fuel use is at the highest volumes, and at zero efficiency at the beginning of acceleration, right when fuel use should be zero. This is the judgement and the conviction of the work equation on present day automotive fuel misuse. This capacity of the work equation to identify the need for energy, only when energy is required will provide the quantitative restrictions on the fixed and moving frames of reference of the Fourth Law and any mechanism it creates.

Ultimately, the combination of both the Fourth and Fifth laws of Motion will demonstrate the alignment of vehicle fuel use with the actual energy required to drive and accelerate the mass of the vehicle. The Fourth Law provides the physical real-world mechanisms of fixed and moving bodies and their interaction, for the proper communication of the load's reaction forces back to the drive train and source of energy. The fifth law of motion provides the theoretical context and control over the physical fixed body/moving body operations so that energy use compliance is guaranteed at all ratios throughout the entire range of speeds. This elevates the fuel use standard over and above today's automotive state of the art, into compliance and alignment. This will only be possible, if a mechanism, a transmission capable of the transmission of mechanical energy in compliance with the requirements of the Fourth and Fifth Law can be designed.

It is the purpose of the next section of this paper, following the conclusion of the science, to explore just such a transmission design.

Conclusion of the Science Section of this Book

The purpose of this discussion on the science of motion was to generate the science needed to power transportation in compliance with the actual energy required to accelerate and drive the mass of the vehicle. The clearer understanding of the action/reaction relationship with the forced and moving load and the supply of power resulted in the development of the Fourth and Fifth Laws of motion. These laws, precisely define the management of the action/reaction relationship between the supply of energy, the engine, and the use of energy, the accelerating load, the vehicle. This energy management between source and load, guarantees compliant fuel use of the vehicle at all times, and aligns vehicle fuel use with the actual energy required to force, accelerate and drive the mass of the vehicle. The quantitative Fifth Law guarantees that this alignment is knowable and sets the standard for the technology needed to bring both the Fourth and Fifth Laws of Motion into reality.

This concludes the science discussion, and introduces the technology section that follows.

PART THREE, THE TRANSMISSION

The Introduction

This is the introduction to the mechanical manifestation of the Fourth Law. The research into the science to develop the needed new laws of motion would have been meaningless without some related mechanical device translating them into reality. The purpose of this opening into the related technology is to explore the obligations and requirements the laws of motion impose on the technology that translates them into the energy compliant world of vehicle transportation. Without a clear understanding of the imposed operating specifications of the Fourth Law on drive line operations, we run the risk of returning back to vehicle fuel use noncompliance which is the automotive standard of today. The intent of this technology is to provide a transmission for vehicle operations capable of all transportation services, yet provide vehicle fuel use that is in alignment with the actual energy required to force, accelerate and drive the mass of the vehicle. Potential fuel use savings would be enormous. Carbon footprint would drop by millions of tons per day, perhaps approaching sustainable balance.

A clear understanding of today's transmissions is essential before transmissions defined by the Fourth Law can be understood and designed. So, what is the function of the transmission in the drive lines of vehicles today? Today's transmissions will be the opening of the discussion in this technology section. The understanding of today's transmissions will reveal the characteristics that make transportation possible, but simultaneously burden all vehicles with non compliant fuel use. Much of this has been covered in the earlier first "Problem" section. The understanding of today's transmissions will reveal their characteristics. These traits will show strengths and weaknesses of today's machines and expose the avenues of further development needed for transmission performance that are in compliance with the new Fourth Law of Motion. The required improvements over and above today's transmissions will then lead to the design of the new transmission that performs according to the Fourth and Fifth Laws. The new transmission design will have to be in compliance with both the Fourth Law and the Fifth Law of mechanical law of energy.

Today's transmission

The purpose of the transmission, all transmissions, is to enable and define the relationship between the driven load and the prime mover that drives that load. Without transmission operations, transportation as we know it, would not be possible. An early transmission, making vehicle motion possible, early in automotive history, was the manual shift four speed. This manual shift four speed will serve as the example for our investigation into today's transmissions. Its simple design serves well for the understanding of the strengths and weaknesses of the transmissions that need to be transcended by the new design.

The clutch is the engagement mechanism that overcomes the speed discrepancy between the prime mover engine and the transmission vehicle load. In order to make the engagement possible, the engine has to have a large mechanical advantage over the first element of the drive line so that vehicle motion due to engagement is possible without burning out the clutch or stalling the engine. The first gear of the transmission provides a high mechanical advantage of the engine over the mass of the vehicle. Then engagement of the clutch is possible, resulting in vehicle motion and acceleration from the standing start. The clutch becomes fully engaged, and the vehicle accelerates with the speed change of the engine. High engine speed with low vehicle speed is the result of first gear. This is very low fuel use efficiency and the next gear ratio defining the next higher fuel use efficiency between the engine and the vehicle is needed. The clutch is disengaged, the transmission is shifted into second gear, clutch re-engaged, engine speed lowered so that the vehicle speed increase can be reapplied. The second gear applies a lower mechanical advantage over the mass of the vehicle, so that the engine speed is lowered, fuel efficiency increased, approaching steady state operating conditions. In a typical four-speed, the gears have to be shifted three times to reach "high" gear for steady state high-way speed and best vehicle fuel efficiency. High "gear" in the typical four speed manual shift transmission usually has a "one to one" input/output ratio. This offers the prime mover no mechanical advantage over the mass of the vehicle, as was available in the first three gear ratios of the transmission.

In high gear, in the typical four speed manual transmission, with a ratio of 1 to 1, the engine rev speed at steady highway speed is the result of the resident drive line ratios and wheel speeds due to tire diameters. Fuel efficiency, in this scenario, is at its best, since the machine based mechanical advantage of the engine over the mass of the vehicle is at its lowest, resulting in maximum vehicle displacement per engine revolution. The engine could not start the acceleration and motion of the vehicle in this high gear ratio, and therefore the transmission has to provide lower, high ratio gears with higher mechanical advantage according to machine theory, for motion from a standing start to be possible.

The gear ratios of the manual shift are relevant and revealing. First gear of a typical manual shift is four to one, input to output ratio. This offers the engine a four to one mechanical advantage over vehicle mass, making motion possible, but at the cost of one quarter the vehicle displacement per engine revolution. This is a fixed ratio, and the only vehicle speed change possible, takes place by speeding up the engine. Engine speed can increase up to max or "red line" requiring the shifting into second gear with a ratio of two to one, cutting engine speed and mechanical advantage in half. This ratio is also fixed and vehicle speed change can only take place, the same as in first gear, by increasing engine speed in the higher transmission ratio, with the lower mechanical advantage. Fuel use decreases as vehicle speed increases according to Graph A discussed in the "Problem" section. When second gear is finished, vehicle speed increases through third gear with a one and a third to one, 1 1/3 to 1, mechanical advantage ratio supplied by the transmission. This ratio is fixed, like all manual shift gear ratios, and vehicle speed increases when engine speed increases. As vehicle speed approaches steady highway speed, the transmission is shifted into high gear, with the highest one to one mechanical equivalent ratio and steady highway speed operations is reached. Engine speed changes, in high gear, can make the changes to vehicle speed needed when in steady state highway speed conditions. When the vehicle needs to be slowed to a stop, the transmission can be downshifted through gears three, two and one if needed. Clutch disengagement and braking, slows vehicle speed to a final stop.

Transmission Characteristics

The principle technical characteristic of automotive transmissions today, is the fixed nature of gear ratio in the progressive sequence of gears ratios in the drive line. First gear is the lowest possible gear, with the highest available mechanical advantage, and no other lower gear is available. The fixed nature of the gear ratios defines the entire range of ratios available for the prime mover, regardless of the fuel efficiency or compliance to the laws of energy. This highly defined relationship between the engine and the load then determines the nature of automotive operations and makes transportation possible. The possibility and benefit of transportation at the individual personal use automotive level, due to fixed ratio transmission drive lines, has outweighed the fuel cost and the related environmental costs. The only possibility of addressing these costs comes from the full understanding of fixed ratio transmission automotive operations and replacement with operations that are Fourth Law and energy law compliant.

The fixed mature of the gear ratios in the transmission translates the inertia of the mass of the vehicle through the transmission, through the clutch engagement device to the engine. The first gear, with its highest mechanical advantage over the load for the engine, makes initial motion from stop possible when the clutch engages. But the moment clutch engagement takes place, forces are applied, via the fixed gears to the vehicle mass load. The reaction forces to these applied forces, are translated, through the drive line, through the fixed ratio gears of the transmission, through the clutch, to the engine, applying a load to the engine, and consequently using energy to apply these forces. The use of energy for this application of force is in noncompliance with the mechanical laws of energy, as has been demonstrated in earlier sections. It is this process of the engagement of sequential, progressive fixed ratio gears that uses energy from the engine for the application of a force, which is in noncompliance with the mechanical laws of motion. This is the essence of fuel misuse in the automotive industry today. Correcting this fuel misuse presents huge potential benefits, and the alignment of auto fuel use with the actual physics-based energy required for auto use purposes.

Related Limitations

The fixed mature of the sequential and progressive gear ratio drive train in the transmission provides one route or path for the flow of forces.

When clutch engagement takes place, forces are applied to the vehicle mass via the gears of the transmission, through the only routing of forces to vehicle mass there is. Then, all the related reaction forces, returning from the forced vehicle mass, flow back to the engine through the same fixed geared transmission and drive line. These reaction forces, load the engine before any vehicle displacement has taken place, forcing engine fuel use for the application of a force to a mass, when, theoretically, no fuel is needed for the application of this force, according to the mechanical law of energy. There is no other power or force routing available through the transmission for the control of vehicle mass and therefore fuel use, misuse, must take place for the application of the automotive drive forces. The singularity of the routing of forces, action and reaction, through automotive fixed ratio transmissions, is the essence of the process that misuses fuel in all automotive transportation today.

This fixed ratio transmission design denies any intermediate access at the transmission for the application of force in the control of the vehicle. Vehicle control takes place via the control of fuel to the engine and the engagement of the clutch. The fixed, sequential and progressive geared ratios are the only ratios available in the acceleration of the vehicle. All vehicle acceleration and drive must come from engine throttle control since no other power control routing or intermediate access of control is possible with any fixed ratio automotive transmission.

The new Fourth Law of Motion, from the previous science section, suggests that the singular routing of the forces, action and reaction, is inadequate for the introduction of energy to the forced motion of vehicle mass. It declares that reaction forces act on fixed bodies and moving bodies, two possible flow routings. No geared, fixed ratio transmissions, are capable of a split power routing from the engine for the acceleration and drive of the mass of a vehicle in the automotive industry today.

The implications of the Fourth Law are far reaching. For the "fixed body" routing of power to be available for the control of vehicle mass acceleration, a mechanism other than the clutch needs to be available for the application of forces. The clutch's sole purpose is the engagement of moving engine components with transmission gears varying in speed from zero to full highway speed. The only conceivable alternative mechanism for the "Fourth Law" based engagement of energy would be ratio based. The "Fourth Law" transmission would have to be able to access the 1 to 0 ratio for the beginning of vehicle acceleration. The meaning of 1 to 0 is that "1" is the symbolic representation of the speed of the engine, and "0" represents vehicle speed. This "1 to 0" ratio would translate none of the reaction forces from the load back to the engine, using no energy from the engine. The reaction forces would be borne by an element of the transmission, most likely the "fixed" case, and the motive force could be applied to the mass of the vehicle without the use of energy, as the laws declare.

This is the essence of the mechanical law of motion. If the fulcrum of a first-class lever is at the very end of the lever, exactly at the location of the load, then any of the motion of the other end of the lever does not do any mechanical work on the load and requires no force or energy. This basic machine theory is the same as the "1 to 0" ratio in a vehicle transmission. In the case of the lever, all of the reaction forces would be borne by the fulcrum and no energy would be needed to move the effort end of the lever. In the case of the "1 to 0" ratio of the transmission, all of the reaction forces resulting from the forced load would be borne by the case even though the load is forced, and no energy would be needed or extracted from the moving, prime mover, source of energy. This is our objective, to apply motive forces onto the stationary, vehicle mass load with the intent to initiate vehicle acceleration, without the use of energy. This compliance with "Work equals Force times Distance," the mechanical law of energy, overcomes the most fundamental barrier of today's automotive technology non-compliant fuel use.

Then, as vehicle motion begins, the "1 to 0" ratio has to vary to something higher so that energy can be extracted from the engine's "moving body" output shaft. As this happens, the vehicle mass reaction forces are shifting away from "fixed body" retention routing to acting on both "fixed" and "moving" bodies routing. The portion of the reaction forces that load the engine output shaft, are the correct energy supply for the vehicle forced motion at the time. This means that the "Fourth Law" compliant transmission is varying its ratio from "1 to 0" toward the higher "1 to 1" ratio according to energy requirements for the drive of the vehicle. The "ones" in the "1 to 1" ratio are symbolic representations of engine speed and vehicle speed, and at "1 to 1", they are the same speed. Reaction force routing at this ratio would be totally onto the engine, it delivering the required energy to drive the vehicle, and the case borne reaction forces would be zero.

The further implication of the Fourth Law of Motion would be the existence of a full range of ratios available for the routing of forces between a source of energy and a forced moving mass. The "1 to 0" to "1 to1" would be the first half of the range of transmission ratios. The second half of the full range of transmission ratios would span from "1 to 1" to "0 to 1." The "1 to 1" ratio, the lowest of the upper range, would be the same as the lower half of ratios, where 1 to 1 symbolically represents input to output of equal speeds. Any increase in ratio of input to output speeds, above "1 to 1," would increase output speeds relative to engine speeds, so that the meaningful ratio definitions, "1 to something," could continue to be used. Ratios of "1 to 2," and "1 to 3," would remain relevant, as engine speed remained steady, and output speed increased. So, a typical upper range, mid range ratio would be 1000 to 2000 rpm, or a ratio of "1 to 2," or 1000 to 3000 rpm, a ratio of "1 to 3." This is consistent, with all the other ratio definitions in the Fourth Law transmission.

Then as the transmission ratio becomes higher, approaching the infinity ratio, this defined ratio nomenclature breaks down. At all of the ratios nearing the infinity ratio, the input shaft speed slows down to less than the defined symbolic "1" speed. The input number would become less and the output number would return to the symbolic "1" representing the output shaft speed, regardless of its actual speed, the moment input shaft speed reaches zero, or stalled. This method of ratio nomenclature provides access to the definition of all of the input/output ratios of the upper half of the range of transmission ratios. Therefore, the highest ratio between input/output shafts of a transmission can then be defined as "0 to 1" and the upper half range of ratios would be from "1 t0 1" to "0 to 1."

So, what does the "0 to 1" input to output transmission ratio mean in the context of the Fourth Law of motion? The "0" of the ratio means that the engine power input shaft would at 0 rpm. The "1" of the ratio means that the drive line for the load would be in motion, and at some speed. This ratio of input/output speeds would be the highest possible ratio of speeds. It would define the upper end of the total range of ratios. Mathematically, any fraction with a divisor of "0," is defined as the "infinity" ratio. It does not mean that the dividend, or the upper number of the fraction is moving at the speed of infinity. The "infinity" term refers to the relationship ratio, when the input shaft speed is zero and the transmission output shaft is in motion. This ratio of speeds, "0 to 1," will be furthermore referred to as the "infinity" ratio. And it will represent the highest limit of the entire range of ratios between two shafts, beginning with the lowest "1 to 0" ratio, and ending with the highest, or infinity ratio "0 to1."

But in the context of the Fourth Law, what does the "infinity ratio" mean. As discussed earlier, energy can only be available in the presence of motion. The power supply shaft at zero speed cannot supply energy to the output shaft load even though it may be fully torqued. (The electric motor can supply full twisting force torque at zero speed.) With no mechanical energy available from the input shaft, the moving output shaft cannot be forced while moving. Its motion and the load will not be bearing any supply energy or providing any load force resistance on the input shaft at the infinity ratio. And since the resistance to rotation cannot be supplied by the output shaft, the action/reaction flow path for the torqued or forced zero speed input shaft has to be between it and the stationary case. This is the reapplication of the secondary route for the action/reaction flow path that was first fully utilized at the "1 to 0" ratio, at first, at the beginning of the range of ratios. This secondary flow path is again fully utilized at the highest ratio at the top end of the range of ratios for the transmission. This requirement of the "Fourth Law" that the "fixed body" be available to bear the reaction forces is the essence of both the first lowest ratio, "1 to 0," and the last highest, infinity ratio, "0 to 1." The technology, the transmission, must have this capability in order to be compliant with the "Fourth Law" of motion, and the mechanical law of energy.

LIMITATIONS OF EXISTING TECHNOLOGY

A clear understanding of the limitations of today's automotive transmission technology provides the barriers that have to be overcome to develop automotive technology and energy use into compliance with the Fourth Law. The knowledge of today's automotive technology will expose all of the reasons for noncompliance of auto fuel use. The understanding of these limitations will reveal their weaknesses in the efficiency of automotive fuel use. These characteristics will then provide the framework of specifications needed to overcome the present limitations, so that technology capable of compliance with the energy laws can be developed for use in automotive transport. The preceding discussion of automotive limitations will assist in the development of the relevant technology needed for compliant automotive fuel use.

Predominantly, there is only one barrier that needs to be overcome so that automotive thermodynamics can align fuel use with the actual energy needed to accelerate and drive the mass of the vehicle. Other restrictions are present, but all are less significant and directly related to the main limitation that blocks auto transport from fuel use compliance. All of the limitations are systemic in nature, and only the systematic application of the new physics, the Fourth Law, and the new technology, will correct the noncompliance of fuel use so that the fuel savings and the related carbon loading reduction can be realized.

Firstly, the fixed ratio transmission is the single most significant automotive characteristic that burdens all auto transportation into non-compliant fuel use. All of the ratios of the sequential and rising ratio transmission gears are each defined as a single ratio in the sequence. Automotive transmissions today have from four to fourteen fixed gears, typically. Though these gear ratios make transportation possible, their predominant restriction of not accessing the lowest ratio or the highest ratio between an input shaft and an output shaft blocks compliance with the laws. Also, the lack of access to all ratios requires that all drivelines have a clutch or some other power engagement mechanism to initiate the acceleration and drive of the mass of the vehicle. This method of power engagement with the load further reinforces the energy misuse from the engine for vehicle motion. The new technology must be fully variable in ratio, and provide a ratio based, or a process based, method for the engagement of force and energy for the motion of the vehicle.

Secondly, an interdependent limitation related to the fixed gear ratio transmission is the single routing of the action/reaction flow path between the engine and the load. The energy flow from engine to load can only pass through the fixed ratio gears of the transmission. This then demands energy from the engine for the application of a motive force in violation of the laws as stated earlier. The clutch couples the engine with the load, applies the motive force on the load, and transfers the load resistance to the engine drawing energy from it, in noncompliance with the laws. The new technology must provide two avenues of energy flow so that compliance to the laws is possible at all the ratios between the input shaft and the output shaft of the transmission. Also, there must be an available interplay between the two energy flow paths so that the correct energy demand from the engine powers the moving load and any other reaction forces are routed to another frame of reference, other than the engine, again, all in compliance with the laws.

Thirdly, the routing of the driveline control inputs must go to the transmission, instead of the engine, as is the standard today. The new technology transmission must engage engine power according to force applied to load, and according to load speed and exercise this engagement with the progressive advancement of ratio of the range of available variable ratios. To do this the vehicle speed control inputs must be routed away from engine control and into transmission ratio control.

The new technology transmission must be able to receive vehicle controls directly in order to vary ratio and process energy use in compliance with the laws.

THE NEW TECHNOLOGY TRANSMISSION PROCESS REQUIREMENTS

The operational process requirements of the new transmission need to be clearly understood, so that the restrictions to energy compliant transportation can be overcome. Understanding the process by which the new transmission operates will help design the new transmission parts configuration. The barriers to compliant transportation, outlined above will provide guidance in the understanding of new transmission process operations.

The new transmission will determine the relationship between the power source and the load, namely the mass of the vehicle. It will receive the input shaft from the engine. The output shaft from the transmission will connect to the drive wheels to power the movement of the load, the mass of the vehicle. These two shafts will operate within a case, and in accordance with the new process that is compliant to the new laws of science. The process by which these shafts operate will be regulated by vehicle operator control inputs sent to and received by the transmission. The transmission must be able to receive and translate all control inputs that result in the desired vehicle motion and do so with compliant energy demands from the source of power. And these requirements must take place at all or any possible ratio of speeds of the input and output shafts.

The control of the shaft speed ratios will be the predominant process of the new transmission. The entire range of ratios, from the lowest to the highest will be fully accessible, and all intermittent ratios are fully homogenous. At the beginning of acceleration, from a standing start, the input shaft will be in motion and the output shaft will be stationary. This is the lowest ratio of the shafts and is the starting ratio, 1 to 0, of the transmission. When control inputs to the transmission, call for the acceleration of the vehicle from standing start, the transmission must be able to apply the motive force to the output shaft in order to drive the wheels. This force application must not require energy from the prime mover. As the vehicle begins to move, the shaft speed ratios vary according to relative engine/vehicle speeds and are fully accessible due to the total homogeneity of the range of ratios. As the ratios increase, the transmission process must draw the correct amount of energy from the prime mover, according to the energy requirements of the moving vehicle. As this motion takes place, full motive forces are still being applied to the mass of the vehicle, but only the portion of the reaction forces reach the engine for the proper energy extraction. The remainder of the reaction forces must be borne by the case of the transmission, a fixed body non-moving foundation.

As the vehicle speed increases, the engine input shaft speed does not change as the output shaft to vehicle drive wheels increase in speed. The force/reaction force management process must continue to load engine and unload case as the energy requirements of the moving vehicle demand. When the input/output ratio reaches 1 to 1 ratio, all of the applied motive force is routed to the engine input shaft and none of the reaction forces are supported by the case. This takes place under the continuous effect of the control inputs to the transmission from the vehicle operator and the full range of variable ratios that the transmission must be able to supply.

When higher ratios are needed, say up to 1:2, for steady highway speeds, the same transmission processes must continue. Operator controls must be routed to the transmission. These control inputs must vary the operating processes according to inputs. The proper drive forces must be applied to the output shaft and to the load. The proper routing of the force/reaction force flow paths must continue so that engine loading and case resistance loading take place for the higher ratios between the input and output shafts. This continues for all transmission ratios.

As stated above, these are the operational process requirements of the new technology transmission. These must take place if the new transmission must operate in accordance with the Fourth Law of Motion. The transmission range requirement from the lowest to the highest ratios verifies the existence of the Fourth Law of Motion. The final phase of the manifestation of the Fourth Law into reality is the schematic layout of the new technology transmission with the description of the process that operates it. There can be no doubt that the schematic is compliant with the Fourth Law. No transmission with a complete range of ratios from the lowest, 1 to 0, to the highest, 0 to 1, ratio is possible without the presence of the Fourth Law of Motion. The transmission and the law reinforce each other's existence.

THE NEW TECHNOLOGY TRANSMISSION SCHEMATIC

This is the technical description of the design of the new transmission. Refer to the schematic 1, in the following document for reference and clarification. As with all transmissions, this new tech transmission has the same basic layout, externally. From the outside, it appears like any mechanical device. Its basic structure is a metal case that contains all the essential internal parts. Connected, but not attached to the case, are two shafts, the input shaft from the prime mover, and the output shaft, connected to the drive wheels of the vehicle. The case and both shafts are visible and labelled in the schematic.

Internally, there are two main operational process hydraulic pumps, unit A and unit B. Each unit is a swash plate hydraulic piston pump. Pump, unit A, is located at the left end, or the input shaft end of the transmission. Pump, unit B is located at the right end, or the output end of the transmission. Between unit B at the right, and the output end of the case, is a proprietary element labelled as the "shuttle sleeve," a critical component in the energy compliant operation of the transmission. Located above the pumps is the hydraulic transfer line that connects units A and B operationally. Located on the lower part of the swash plates of both units, are the control input tabs, A and B. A line to the hydraulic oil sump is supplied for each pump unit, A and B.

Hydraulic pump, unit A at the input, left end of the transmission is installed according to standard hydraulic operational protocols. The piston block outlets seal against the transmission case. The pistons reciprocate in and out under the action of the swash plate. The swash plate is made from two plates, one plate rotates, and the second plate pivots, but does not rotate. The rotating plate attaches to the pistons and forces them to reciprocate. The second fixed swash plate is held stationary by pivots and a pivot frame. The pivot of unit A, is held rigidly in place to the transmission case. The stationary swash plate tilts on the pivot to the desired angles, taking the rotating swash plate with it. The changing angle of the rotating swash plate changes the stroke length of the reciprocating pistons, which changes the volume of oil being pumped by the unit. Oil flow passes from the outlet of the piston block into the proper two chambers housed in the end of the case. One oil chamber is routed to the transfer line for operations with unit B. The other chamber is routed to the sump line for oil supply or return operations to the sump. Control inputs are applied at the bottom of the fixed and pivoting swash plate to angle the plate and vary oil flow as required.

Hydraulic pump, unit B, at the output, right end of the case, is similar to unit A, but is installed in a unique manner. Unit B retains the piston block, pistons, swash plates, moving and pivoted, pivot and pivot support, the same as unit A. Unit B has its own separate control inputs at the swash plate tab. But the most significant difference of unit

B is that the oil flow from the piston block does not pass directly to case and transfer line as in unit A. Instead, oil flow from and to, the B piston block enters the shuttle sleeve. The shuttle sleeve is a proprietary element designed to route oil flow from the piston block to the proper ports, high and low pressure, regardless of the orientation of the swash plate to the case. This is necessary because unit B swash plates rotate as a unit as part of the transmission process. The shuttle sleeve element is designed so that oil flow from the unit always flows to the correct ports and lines. This orientation guarantees that there is a constant process interaction between the hydraulic units at any time of the transmission process.

The shuttle sleeve, labelled in the schematic, is a mechanical disc that is the same diameter as piston block B. A projection from the center of the disc passes through the case at the output end and connects to the output shaft. Oil flow routing through the disc and the projection guarantees uninterrupted transmission processes between the hydraulic units. This is necessary because the unit B swash plates, the pivot, and pivot supports, the shuttle sleeve and output shaft rotate as one, as a single unit. The shuttle sleeve then shuttles the hydraulic oil to the proper ports, regardless of the orientation of the swash plates to the piston block of unit B. Process oil is shuttled to the high-pressure port, the transfer line, and unit A. Supply and return oil is shuttled to the low-pressure port and the line to sump. This customized installation of unit B is necessary so that the hydraulic/mechanical dynamics needed to transmit power, according to the Fourth Law are possible at any speed ratio between the input and output shafts. Details of the process are outlined in the following documents.

This transmission is the manifestation of the Fourth Law of motion into the physical world. It is the transformation of the invisible into the visible. The schematics illustrate this transmission. It is a metal case with input and output shafts entering and exiting at opposite ends. The input shaft connects with both piston blocks of both hydraulic units A and B and nothing else. Both piston blocks rotate with the input shaft as one. The output shaft connects with the shuttle sleeve, and unit B swash plate via the pivots and pivot support. These elements rotate with the output shaft as a unit. Pistons reciprocate in and out of the blocks according to block rotation and swash plate angles, pumping and receiving process hydraulic oil to the opposite unit via the transfer line. The swash plate angles govern the oil flow and speed ratio of the shafts according to the transmission control inputs. Energy is taken from the input shaft prime mover according to energy demand by the output shaft moving the load.

Actions and reactions of the drive forces move through the transmission according to the Fourth Law, and energy use is aligned with the actual energy required to drive the output shaft load. The following documents and schematics describe and illustrate process operations at various ratios from the lowest to the highest. Compliance to the energy law and the Fourth Law of motion is demonstrated.

Hydra-Mechanical Transmission at 1 to 0 Ratio Schematic 1

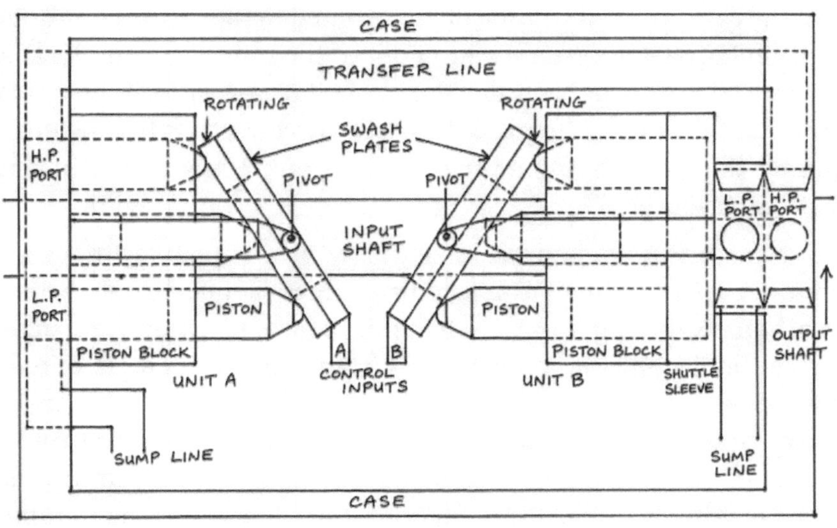

This is the schematic diagram of the Fourth Law compliant transmission at the lowest ratio of 1:0. The "1" symbolizes the moving input shaft. The input shaft would be moving at the prime mover's designated speed. It would be rotating in the clockwise direction. This means that the piston block surfaces facing the viewer are moving in the downward direction. Both piston blocks are moving clockwise and downward since both are moving together with the input shaft. The "0" symbolizes the zero speed of the output shaft, exiting the right end of the case.

The swash plates at this 1:0 ratio, are at maximum angle. This means that, as the piston blocks move, the pistons are reciprocating the full length of their travel or stroke. Pistons move in and out of the block as they reciprocate under the control of the swash plates. The full stroke movement of the pistons, pumps the maximum volume of hydraulic oil as they move. The only components that are moving at this ratio configuration are the piston blocks, the pistons and the oil flow that the pistons are pumping.

In this 1 to 0, configuration, oil flow through units A and B is at maximum. Unit B pumps the oil from the reservoir, through the shuttle sleeve, through the transfer line, through unit A and then back to reservoir. Unit B at the right end of the case, draws in the full volume of oil from the sump through the sump line. Unit B pistons, on the near side of center, that are reciprocating out of the block are drawing the oil into the cylinder. Then as the block rotates, and the pistons reciprocate into the block, on the far side of center, the oil is pushed out of the cylinders, through the shuttle sleeve and into the transfer line to unit A at the left end of the case. The full volume of oil is routed to the pistons on the near side of center, that are reciprocating out of the unit A piston block, and are capable of receiving the full volume of oil from unit B. Then as block A continues to rotate, the pistons on the far side of center, reciprocate back into the block, pushing the oil out the sump line and returning it back to reservoir.

No transmission action has taken place at this point without input from the controls. When the first control force is applied to tab A on unit A swash plate, forcing it to the left, all of the transmission force actions and reactions begin. The control force on tab A, to the left tends to reduce swash plate angle from maximum angle toward perpendicular. The swash plate angle does not change but the application of the control force tends to reduce the oil flow volume of unit A. This action immediately increases hydraulic oil pressure on both units A and B. The oil pumped from unit B is positive displacement in nature, and any restriction of the flow immediately increases the hydraulic oil pressure on unit B but also on unit A. This increase in oil pressure begins the action reaction forces on the shafts so that transmission action can take place.

Control inputs on tab A result in high pressure oil being pumped from B to A. The clockwise motion on block B pumping at high pressure applies action/reaction forces on two elements of the transmission. Pumping forces are being applied to the input shaft and forces are also being applied to swash plate B. The input shaft is driving block B and this pumping action requires energy from the input shaft and applies a resistance to rotation or a load on the input shaft. As the input shaft drives block B, the angled awash plate forces the pistons into the block, pumping the oil out, against high pressure. The reaction force of these pistons onto the swash plate is a torque on the angled swash plate in the clockwise direction. Unit B swash plate is connected to the output shaft, and this torque is being applied to the output shaft and to the vehicle drive wheels in the desired clockwise direction. This demonstrates how the control force on tab A, results in the motive drive force being applied on the output shaft load.

Understanding the force action/reaction process at unit A, the other hydraulic unit, completes the energy use illustration in this schematic and at this ratio. High pressure, positive displacement oil is being driven into the receding piston cylinders forced to receive the full volume of oil. The force of the pistons being driven out of block A cylinders applies lateral forces on two elements of unit A. Piston forces are being applied to swash plate A, and reactions to these forces are driving block A in the clockwise direction. The forces of the pistons being pushed out of the block cylinders against swash plate A apply a counter-clockwise torque on A unit swash plates. This then applies a counter-clockwise force on the case to which swash plate A is attached. The counter-clockwise case reaction forces are "equal and opposite" to the driving forces applied to unit A piston block. These reaction forces provide the required case counter clockwise reactions to the drive force applied to the output shaft.

The second reaction force of the high-pressure pistons of unit A is the driving action of the piston block A in the clockwise direction, the same direction as input shaft motion. This reaction of piston block A driving the input shaft in its direction of motion balances the load force applied to the input shaft at block B when pumping the oil at the beginning of the process. As a result, the net force of piston blocks A and B on the input shaft is zero. And the input shaft moves as intended, drawing no energy from the engine, even though the output shaft, at zero speed, is fully forced in the intended clockwise direction of motion, with the case bearing all equal and opposite reactions forces. The input shaft is not loaded and the output shaft is fully loaded with the intended drive force.

The mechanism above, schematic one, shows the transmission, able to process transmission forces in full compliance with the laws of motion. Any control force applied to tab A, pushing to the left and tending to move swash plate A toward the perpendicular, initiates an increase in hydraulic oil pressure. This oil pressure varies directly with the control force pressure applied to tab A. The high-pressure oil is pumped from unit B piston block to unit A piston block. B block loads the input shaft and A block drives or motors the input shaft with equal energy as B, so that net load on the input prime mover is zero. The reaction forces of the pumped oil at unit B is the forcing of swash plate B and ultimately the output shaft in the clockwise direction, which was the original intent of the control force at tab A. High-pressure oil forces at unit A, force the swash plates and consequently the transmission case in the counter clockwise direction. The fixed case provides the foundation for the reaction to the force applied to motoring piston block A, that balances the load on the input shaft, taking no energy from the engine. The fixed case carries the reaction force of the fully forced, non-moving output shaft, in compliance with the laws of motion.

Compliance to the laws of motion, including the Fourth, and the mechanical law of energy, is complete. The application of the motive force on the output shaft, with the intent to start motion takes no energy from the input shaft or prime mover. This is in complete compliance with, "Work equals Force times Distance," the mechanical law of energy. Every other transmission uses energy to apply the motive force on the load and is in violation to this mechanical law of energy. The reaction forces to the forced, non-moving output shaft are applied to the non-moving case via hydraulics. This complies with the Third, "Every action has an equal and opposite reaction," and the Fourth, "Equal and opposite reactions act upon fixed bodies," (abbreviated). The transmission in above schematic 1, is in compliance with the law of energy and the laws of motion. And as such, it overcomes all the barriers to the correction of automotive fuel misuse.

Documents that follow, illustrate other ratios in the transmission process, and the related compliances to the laws of energy and motion.

Hydra-Mechanical Transmission at 1 to .5 Ratio Schematic 2

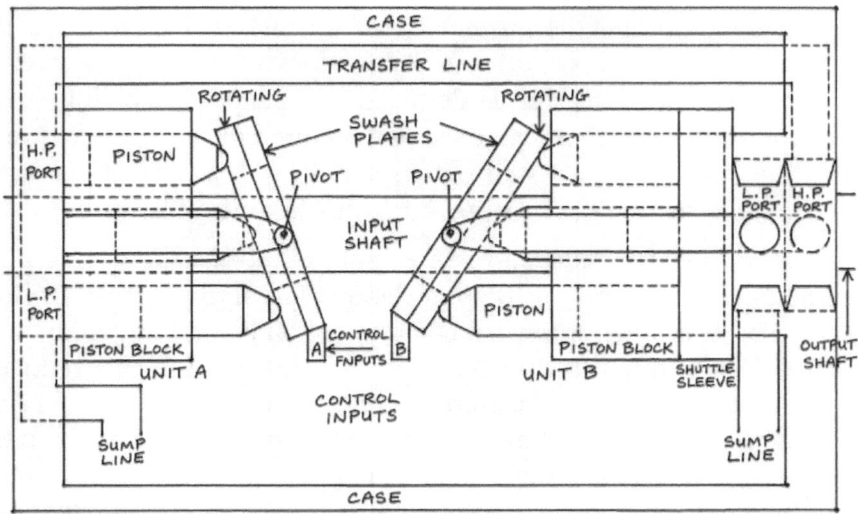

Schematic 1, previous, showed all of the transmission actions at the 1:0 ratio, with a drive force being applied to the output shaft load so that the vehicle load could start to move. Schematic 2, above, shows the transmission action, energy use and force flow, after the vehicle has started to move, at a transmission ratio of 1:1/2 and under the influence of the applied force. As before, the "1" symbolizes the motion of the input shaft at any designated, design speed and in the clockwise direction. The output shaft, or vehicle load is moving at half, .5 or 1/2, the speed of the input shaft speed. It is moving at half speed under the influence of the drive force, and therefore requires energy from the prime mover engine to do this. Schematic 2 shows how this process works, all in compliance with the laws.

The output shaft in this schematic is rotating clockwise, in the same direction as the input shaft. It is moving at half the speed of the input shaft speed. Since the input shaft speed has remained the same, the motion of the output shaft at half speed changes the relative motion of the internal components of the transmission. The moving output shaft means that B swash plates, unit B pivots and supports, the shuttle sleeve, and the output shaft, namely, the swash plate output cluster, is moving as a unit, at half speed in the clockwise direction. Piston block B is at speed 1 and swash plate B, output cluster is at 1/2 speed, clockwise. This means that the relative motion between B swash plates and B piston block has been reduce to half of what it was in the previous schematic at ratio of 1:0.

The swash plate, output shaft cluster includes the shuttle sleeve unit. The moving swash plate, pivots and output constantly change the relative position and motion of the cluster to two transmission components. The cluster constantly moves relative to piston block B and the transmission case. The purpose of the shuttle sleeve, is to route process hydraulic oil, to the correct designation regardless of its motion or position relative to block B or the case. The piston outlets are always routed to the correct port for sump suction or high-pressure, transfer line oil flow.

So, what does the reduced relative motion between the swash plate and the piston block of unit B do to the transmission process? The half speed, relative motion reduces the volume of oil being pumped by unit B, even though the swash plate angle is at maximum. This reduced, positive displacement of oil is transferred to unit A via the transfer line. The reduced oil flow would lead to a drop in hydraulic oil pressure, if the oil volume in unit A was not reduced by the same volume. The control force applied to tab A, on unit A swash plate moves the plates so that unit A volume is half of maximum. The plate is more perpendicular at half of maximum angle, able to receive half oil volume and thereby maintain process oil pressure. Because of the reduced volume received by unit A, the oil pressure remains under control and varies directly with control force pressure, an operational requirement of the transmission. We see that process operations are possible in the transmission at the 1 to 1/2 input/ output ratio. The transmission in schematic 2, at the ratio of 1 to 1/2, and its interplay with the laws of motion and energy, will broaden the understanding of transmission operations and the laws that govern them.

Energy needs to be extracted from the input shaft in order to drive the loaded output shaft at half speed. Unit B piston block drives the output shaft via the shuttle sleeve cluster. There are two sources of load applying resistance to rotation onto the input shaft via block B. Half a work load unit is being carried by block B by driving the loaded output shaft at full load, but at half speed. The second work load unit is being applied to block B by pumping half volume of hydraulic oil at full pressure to unit A. These two sources of load place a full work load unit on block B and the input shaft.

The energy picture is not yet complete since the input shaft is fully loaded at block B, even though the output shaft is only moving at half speed. The full speed input shaft should only be loaded at half load.

This correction of energy use takes place at hydraulic unit A. Unit A receives half volume of positive displacement pressurized oil from block B via the transfer line. Process pressure is maintained by reducing the angle of the swash plates, from maximum, to the more perpendicular angle able to control piston displacement to half volume. This reduced swash plate angle and the reduced piston stroke is visible in schematic 2. The more perpendicular angle, the reduced angle of the swash plate A reduces the lateral reaction forces that the full oil pressure applies onto swash plate A and piston block A. The pistons at unit A apply the same force as in schematic 1, but the angle here is not at max. The reduced angle of swash plate A reduces the lateral forces of the pistons on block A and on swash plate A. These lateral forces are half of what they were in schematic 1. These half lateral forces torque the swash plate counter clockwise and place a half counter clockwise force on the case. The piston half lateral reaction forces drive the piston block in the clock wise direction, reducing input shaft full load from unit B by half. This corrects the input shaft loading to half load, what it should be while driving the fully loaded output shaft at half speed. The full speed, half load input shaft energy equals the full load, half speed output shaft, as per laws. Proportionally, half of the reaction forces are borne by the case, and half are carried by the input shaft as required by the Fourth Law.

Third and Fourth Laws of Motion

A key characteristic in the relationship between the third and fourth laws of motion becomes evident in schematic 2. The third law requires an equal and opposite reaction force from the force applied to the moving output shaft. These reaction forces are carried by piston block B and are converted into two flow paths of energy and force. These two force flows are mechanical and hydraulic. The mechanical energy is carried by the input shaft and the hydraulic energy is carried by oil flow to hydraulic unit A. The mechanical reaction force loads the input shaft. At unit A, the oil pressure and flow are converted to an action/reaction relationship between the input shaft and the fixed transmission case. The hydraulic force applied to the moving block A, is equal and opposite to the reaction force applied to A swash plate and transmission case. These forces at A use action/reaction dynamics to extract the correct amount of energy from the prime mover given the force and the speed of the output shaft. The energy extraction process at A, translates the correct reaction forces onto the fixed case, a continuous interaction. These reaction forces can be fully hydraulic, fully mechanical or a proportional combination of both hydraulic and mechanical energy, as is required by the Fourth Law.

The fuller understanding of the Third Law is that it sets the stage for the Fourth Law extraction of energy needed for forced motion to take place. The Third Law provides the reaction force from the applied force on the output shaft. This reaction force is not applied to a single support such as the case. The reaction forces of the Third Law provide the information for the extraction of the proper amount of energy from the prime mover. At many ratios the load force needed to torque the engine, to make it provide energy and power, needs to have the case of the transmission provide the reaction force for the engine torque to take place. This can only happen with the Third Law reaction forces being used by the mechanism, to act upon fixed bodies, the case, moving bodies, the input shaft, or proportionally both, the conditions of the Fourth Law. Without Third Law reaction forces being available and the Fourth Law assigning them to their correct application, the proper use of energy would not be possible in driving a forced, moving body. The Third Law's supply of reaction forces and the Fourth Law's application of the reaction forces on the case and/or the input shaft, vary the energy removed from the prime mover to exactly match the energy needed to drive the output shaft load. This is done homogeneously, at every incremental change of input/output ratio or load force.

At the risk of overstating it, the sole purpose of the Third Law of motion is to provide the data to the Fourth Law of motion so that it can calculate and extract the correct energy needed from the prime mover to power the forced moving load. Finally, energy is part of the laws of motion.

Schematic 3 that follows, illustrates transmission operations at the 1:1 input/output ratio.

Hydra-Mechanical Transmission at 1 to 1 Ratio
Schematic 3

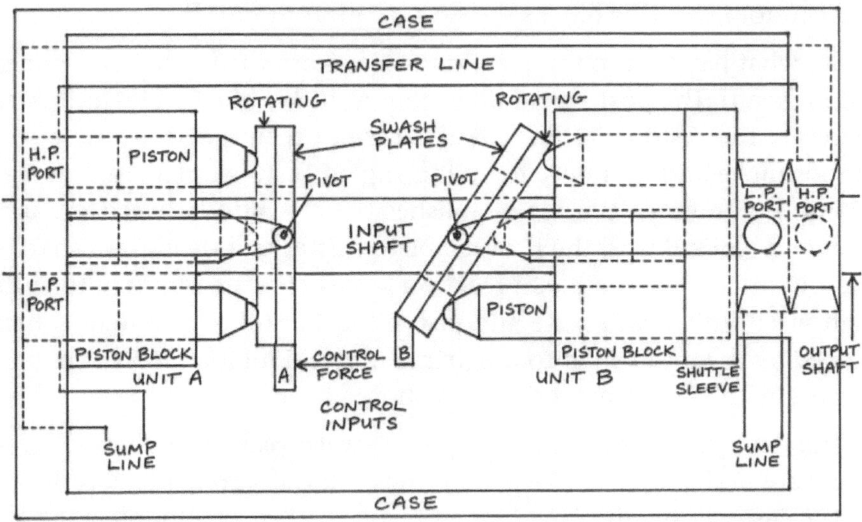

Schematic 3 shows transmission operations at the 1:1 input/output ratio. Standard operations have advanced from the 1:1/2 ratio in the previous schematic to this higher ratio under the influence of the control force at tab A. The control force at A is pushing the tab to the left and is tending to move swash plate A toward a perpendicular angle to the input shaft. Being square to the input shaft is merely a reference and has nothing to do with operations, directly.

Standard transmission operations have not changed. The input shaft, entering the case at the left and connecting to both piston blocks, A and B, is moving clockwise, with block surfaces nearest the viewer moving down. Unit B swash angle is still at maximum, and being held at maximum by control forces on tab B. Piston block B is driving the output shaft clockwise via B swash plate, pivots, shuttle sleeve and output. Output shaft load force reaction forces are translated back to the input shaft via block B. The hydraulic oil from B, is being transferred back to unit A, and is pressurized according to output load reaction forces and swash plate B maximum angle.

One operating condition has changed. The hydraulic oil flow has stopped. The 1:1 input/output ratio requires that the swash plate A be perpendicular to the input shaft, or square with piston block A. At this position, the pistons in block A have stopped reciprocating, even though piston block A is still moving clockwise with the input shaft. Since the pistons no longer stroke in and out of the cylinders, unit A cannot receive any oil from unit B. The oil from unit B is deadheaded at unit A, and cannot flow. This blocks the flow of oil from unit B.

So, what happens to the reaction forces from the loaded and moving output shaft? The positive oil flow from unit B is blocked by unit A and therefore the pistons in block B also cannot reciprocate. These pistons cannot move and therefore force the angled B swash plate to move in a 1:1 ratio with block B. This establishes the 1:1 ratio between the input shaft and block B with the B swash plate cluster and the output shaft. As a result, full output shaft load reaction forces are transferred to the input shaft via B unit swash plate and piston block. Full output load is being carried by the input shaft to the prime mover. Full load forces are being carried by the prime mover, as they should be at the 1 to 1 ratio.

The corollary of the square angle of swash plate A is that no lateral forces are available from the pistons in block A, to drive block A or react on the case via swash plate A. As a result, Unit A cannot alter the load on the input shaft, to modify load forces from unit B. And there are no reaction forces on the case. All the reaction forces from the output load are mechanical and act upon the moving body, input shaft, as the Fourth Law forecasts. The extraction of energy from the prime mover is equal to one load torque, equal to the reaction load from the output shaft with both moving at the same speed. Energy from the prime mover is equal to the energy needed to drive the output load, the correct calculation from the Fourth Law, and in compliance with the laws of energy.

The Fourth Law of Motion imposes two flow paths for energy transfer between the input and output shafts, hydraulic and mechanical. At this 1:1 ratio, the hydraulic oil flow path is pressurized, but non-flowing. This process statis transfers all input/output energy transmission via the input shaft, the mechanical path for energy flow. The non-flowing hydraulic oil ensures that the case of the transmission is not at all loaded with any reaction forces from the output at this ratio.

Schematic 4 that follows, illustrates transmission operations at the 1:2 input/output ratio.

Hydra-Mechanical Transmission at 1 to 2 Ratio Schematic 4

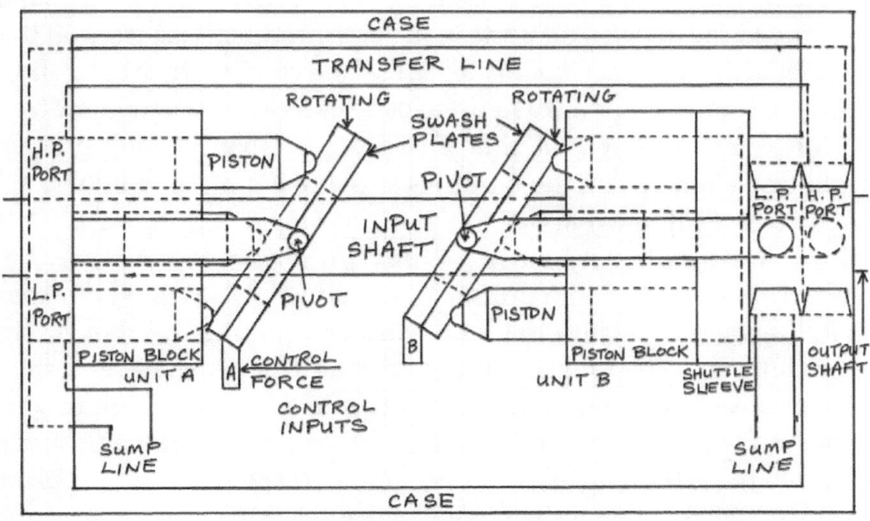

Schematic 4 illustrates the Fourth Law transmission in the 1:2 input/output ratio. Basic operations remain the same as in the previous schematics. The input shaft enters into the transmission case from the left and connects to both piston blocks, A and B. The input shaft is connected to nothing else in the transmission. It rotates in the clockwise direction, moving the near surfaces of the blocks in the downward direction. Swash plate B, the pivots, the shuttle sleeve and output shaft are one unit and also move in the clockwise direction. Control forces are still applied to swash plate tabs A and B. Swash plate B and tab B are being controlled by control forces to remain stable at the maximum angle as in the previous schematics. Control forces applied to tab A, have moved swash plate A angle to maximum, opposite to max angle in the first schematic at ratio 1:0. The control forces cannot move tab A any further to the left, as they have reached the limit of travel, on swash plate A.

This maximum angle of swash plate A changes hydraulic operations from its previous perpendicular position at ratio 1:1 in schematic 3. The pistons in block A reciprocate at full stroke and are able to process the full volume of positive displacement hydraulic oil. But the hydraulic oil flow is reversed from that of the early ratios. The piston travel in the cylinders is in the opposite direction as when the swash plate angle was,

"

say, positive, as in the schematics 1 and 2. Piston block A is moving clockwise downward, and swash plate A at this angle is pulling pistons out of the cylinders on the far side of center and pushing pistons into the cylinders on the near side of center. Hydraulic oil is drawn into block A at the far side of center and pushed out from the near side of center according to the internal oil ports in the case head. This reverses oil flow from that of the earlier ratios, 1:0 and 1:1/2, and converts hydraulic unit A into a pump and unit B into a motor. Positive displacement oil flow from unit A is routed through the shuttle sleeve, such that the high-pressure oil is always aligned with the B pistons stroking out of the cylinders and able to receive full oil volume. As the pistons push out, they act on the angled B unit swash plate, moving it forward relative to the block. And since the oil volume of both units is equal, swash plate B moves forward one full revolution relative to block B, resulting in the speed of two on the output shaft. The input shaft provides one output shaft revolution, mechanically, and system hydraulics provide the second output shaft revolution hydraulically. The reaction forces to these dynamics, will factor into the extraction of the proper amount of energy from the prime mover to drive the output shaft work load at twice the input shaft speed.

As before, all the output load reaction forces are being carried by unit B. Unit B swash plate is driving the output shaft load force and the related reaction force is transferred back to block B via lateral piston forces from the swash plate. Block B processes two reaction flow paths, mechanical and hydraulic. The mechanical reaction force from driving the output shaft is transferred back to block B via the angled B unit swash plate. The lateral forces that the angled swash plate applies to the block transfers the reaction forces to the block and the input shaft. The reaction force mechanical load on the input shaft, at unit B, is one output shaft load force.

But the output shaft is moving at double the speed of the input shaft. This means that the hydraulic reaction force must add one more load force onto the input shaft at unit A. Hydraulic oil pressure defined at unit B as a function of output load reaction forces is applied to the pistons by the angled swash plate. This defines unit A discharge pressure due to the positive displacement nature of the transfer of the pumped oil. The one load force oil pressure at unit A creates piston resistance to stroke as they push the oil out of the block on the near side of block A center line. The lateral forces of the pressurized piston stroke create two load force reaction forces. These reaction forces act on unit A swash plate and unit A piston block.

At swash plate A, the stroke resistance of the pistons as they pump pressurized oil to unit B applies a lateral reaction force onto the swash plate in the clockwise direction. This applies a load force reaction onto the case of the transmission in the clockwise direction. This applied lateral reaction force on the swash plate requires an equal load force resistance to rotation of piston block A. This load force reaction force on block A is in the counter clockwise direction, loading the input shaft with one load force. As a result, the input shaft that drives both piston blocks is loaded with two output shaft load forces, one mechanical from block B and one hydraulic from piston block A. The input/output energy relationship of the two load-force loading of the single speed input shaft is equal to the one load force loading of the output shaft at double speed. The input shaft supplies the two load forces that the output shaft needs to be driven at one load force, double speed.

Noteworthy at this ratio is the relationship and the proportionality of the fixed and moving bodies, the engine and the case. For the constant speed engine to be able to provide double work load energy for the double speed output shaft, it must drive a torque load higher than the reaction forces of the driven load. The single load reaction forces of the output shaft would not be able to provide the higher resistances to rotation needed to load the engine for the higher energy output. The higher torque needed to load the engine comes from the proportional relationship between the engine and the case. The reaction of the speed ratio places the resistance to rotation for higher engine torque onto the case, the fixed body. The case provides the resistance to rotation of the engine needed for the extraction of more power for the higher input output ratios. The case lightens the input load at the ratios lower than 1 to 1 and increases the input load at the ratios higher than 1 to 1. All this takes place in compliance with the Fourth Law.

The proportionality requirement of the Fourth Law is essentially the application of machine theory to the power transmission process. The source of power from the prime mover would be a constant speed output shaft able to deliver variable levels of torque or load. The vehicle load would be variable in speed but constant in load force. Input/output shaft speeds are constant vs variable. Input/output shaft loadings are variable vs constant. Shaft speeds are constant vs variable. Shaft loadings are variable vs constant.

At the beginning of the transmission process the drive force is applied to the mass of the vehicle. At this point all the reaction forces of the applied load force are being carried by the fixed body transmission case.

Then under the influence of the drive force the vehicle starts to move. At the instant vehicle motion takes place, the shaft ratio changes and machine theory comes into play. If the transmission were all gears the gear ratio at very slow vehicle speed would be say, 50 to 1 input to output. This machine ratio would then load the engine just enough to drive the vehicle at the slow speed. But the Fourth Law schematic has no gears and still has to load the engine just enough for the proper energy to drive the slow speed vehicle. This parallel function to machine theory is being carried out by the schematic proportionality between the fixed case and the moving engine output shaft. At slow vehicle speed, just enough of the load reaction force has been shifted away from the case and onto the engine output shaft to provide energy for motion. This proportional loading is exactly the same as machine theory and is possible with the absence of gears.

The machine theory, proportional dynamic is always at play at all ratios other than 1 to 1. At slow speed lower ratios, the fixed case carries the portion of the load reaction forces that would be too much for the engine energy delivery. Then at fast speed high ratios the proportional function becomes more significant. At high ratios the loading of the engine has to be with forces higher than the reaction forces returning from the load. In the presence of gears the higher ratio would generate the higher engine load required for the extraction of the needed energy. The action/reaction relationship of the fixed case and the moving engine shaft generates the higher forces needed for the supply of energy. The case provides resistance to engine rotation at all forces needed for energy supply, above 1 to 1 right up to maximum torque engine stall. Only the gearless schematics in this section would be capable of the full range of engine loadings needed for full range input/output ratio operations.

The input/output ratio dynamic is in full compliance with machine theory, but is completely accessible through gearless proportional schematics controlled by the Fourth Law. The placement and generation of the related reaction forces of the proportional theory transmission of power runs parallel to machine theory transmission of power. Fourth Law proportional power process is the same as a machine theory transmission with an infinite number of gear sets providing the ratios for the entire range of input/output ratios. Such a mechanism would be impossible to construct but the Fourth Law proportionality schematic that parallels machine theory power transmission is outlined in these schematics and is easily accessible.

Essentially, the Fourth Law transmission is a mechanical power transformer. Any action of the swash plate controls results in the application of force and displacement needed for the drive of the load and the flow of power from the prime mover to that load. The power flow from the engine uses a single speed variable force supply of energy and transforms that energy into the variable speed, steady load force power needed to drive the load. At all input/output ratios, this transformation process takes place so that the output always gets the power it needs to move the load. Only the 1 to 1 ratio and the infinity, 0 to1 ratio are of note in this transformer process. At the 1 to 1 ratio, the transformation of energy does not take place since the speed of both shafts and the load of both shafts is the same. Power still flows but transformation of that power is not needed at the 1 to 1 ratio. At the infinity ratio, no power flow takes place, so the transformation process of power stops. At all other ratios in the entire range of ratios from 1 to 0 to 0 to 1, the power transformation process takes place taking energy from the source and applying it to the load, all in compliance with the laws of energy and motion.

Schematic 5 that follows, illustrates Fourth Law transmission operations at the 1:3 input\output ratio.

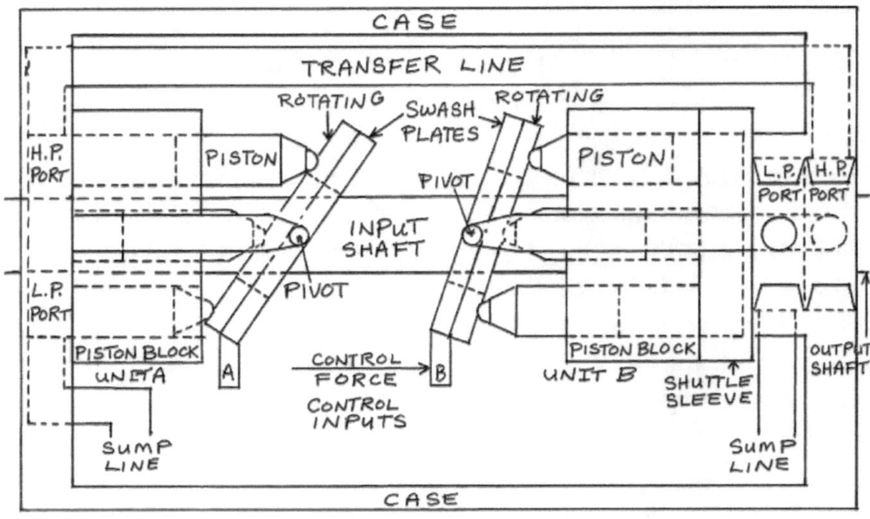

Schematic 5 illustrates the Fourth Law transmission operations through ratios above 1:2, reaching to ratio 1:3. Operations remain similar to earlier schematics, except for the application of control forces. This range of ratios is under the control of swash plate B, and the change of plate angles due to control force inputs.

Basic transmission operations remain the same. The input shaft enters into the transmission case from the left, and attaches to both piston blocks A and B, and nothing else. Input shaft motion is clockwise, as is the motion of both piston blocks. The output shaft exits out of the transmission case at the right. Internally the output shaft is connected to the shuttle sleeve, pivot braces and pivots, and B swash plate. These elements rotate with the output shaft as one unit in the clockwise direction of rotation, the same as the input shaft. Control forces are applied to tabs A and B. Control forces at tab A, at this ratio make no angle change but only serve to stabilize swash plate A at this maximum angle. The control forces at tab B serve to change the angle of swash plate B from maximum angle, as in schematic 4 at 1:2 ratio, to half volume angle at 1:3 ratio in schematic 5, above.

Transmission operations achieve the 1:3 input/output ratio using both mechanical and hydraulic processes. Piston block B, which is carrying one reaction load force of output shaft drive, transfers this resistance to rotation to the input shaft. This mechanical transfer of load resistance to the input shaft, draws one load force of energy from the prime mover. This mechanical rotation of block B drives the output shaft 1 clockwise revolution of the three revs total output. Hydraulic operations account for the remaining two rotations of the output shaft, and the related loading of the input shaft and energy use from the engine.

Hydraulic operations under the influence of the changing angle of swash plate B, result in the increase in oil pressure over and above the earlier ratios. Hydraulic oil pressure in all previous ratios and swash plate angles from schematic 1 (ratio 1:0) to schematic 4 (ratio 1:2), have remained relatively unchanged. Process oil pressure in the previous ratios was a function of piston block B driving swash plate B, at maximum angle, against the reaction load forces of the output shaft. The swash plate angle applies the downward force on the pistons, that results in the operating hydraulic oil pressure. Until this ratio of 1 to 3, hydraulic oil pressure remains stable, and has remained an unchanged characteristic of reactions to output load force. Now, any ratios higher than 1 to 2, result in increased operating hydraulic pressures than in lower ratios.

The application of the control forces on tab B result in the higher input/output ratios that increase operating hydraulic oil pressure. One full volume of oil is pumped from unit A, at maximum swash plate angle, for every piston block A and input shaft revolution. Unit B was able to receive this full volume of oil when its swash plate angle was maximum and oil volume was equal to unit A. But ever since the 1:2 ratio, in schematic 4, the control forces on tab B have been pushing to the right and reducing swash plate B angle from max toward the perpendicular. This plate angle change tends to reduce unit B oil processing volume. Due to the positive displacement nature of oil transfer from unit A, the reduced volume of unit B results in increased operating oil pressure.

At the illustrated swash plate angle, in schematic 5 above, oil volume of unit A is double the volume of unit B. The higher volume of positive displacement oil is received by unit B, via the shuttle sleeve, that routes the oil to the pistons of B that are reciprocating out of the block. This results in clockwise motion of swash plate B relative to the piston block B. When the relative motion of B swash plate to B piston block is two clockwise rotations, unit B is able to receive a full volume of hydraulic oil from unit A. These two hydraulic clockwise rotations of swash plate B are added to one mechanical clockwise rotation of piston block B that results in three output shaft revolutions for every one input shaft rev, the desired 1:3 operational ratio.

The shallower, more perpendicular angle of unit B swash plate results in higher reaction forces on the pistons in unit B. The pistons that apply the lateral forces on B swash plate have to apply a higher force onto the plate to overcome output shaft reaction forces, which remain at one reaction load force. These half volume, shorter piston stroke forces need to be twice that of earlier ratios, resulting in twice the operating oil pressure being pumped by unit A. Unit A, pumping at full volume, then has two times the piston lateral forces applied to piston block A, and swash plate A. The resulting doubling of reaction forces on swash plate A, place a clockwise rotation force on the fixed case. The same, doubled, lateral piston reaction forces apply an equal and opposite counter clockwise force on block A, loading the input shaft with two output shaft load forces. The fixed case provides the necessary resistance to rotation reactions for proper energy extraction loading of the prime mover at this ratio. Force action/reaction dynamics between the fixed case and the moving A piston block and input shaft are in place by the hydraulics to take the correct amount of energy from the prime mover at this ratio.

We see that the input/output energy balance is in compliance with the laws. The output shaft is driving a single load force at a speed three times faster than the input shaft speed. The reaction load force remains at one unit even though the output speed ratio is 1:3. The single load force reaction force to driving the load is being carried by hydraulic unit B. Unit B processes this reaction force according to the Fourth Law into two energy flow paths, one mechanical and one hydraulic. Mechanically, a single reaction load force is being transferred by block B onto the input shaft to take the energy from the prime mover needed for one of the three output revs. Hydraulically, the control force on tab B swash plate has moved the angle of the plate, away from maximum to the shallower, more perpendicular, half oil volume illustrated above. This plate angle, halves the oil processing capacity, and doubles the pressure. At half volume, unit B swash plate and output shaft, must rotate two times relative to block B to be able to receive the full single rev volume from unit A. As a result, the output shaft speed is three. At double the oil pressure in unit A, the lateral piston forces on the swash plate and block A are doubled. The unit A swash plate doubled torque is applied to the transmission case in the clockwise direction. The reaction to this transmission case force is the same applied torque on block A, counter clockwise loading the input shaft prime mover with two load forces, additional to the mechanical single load force from unit B. Prime mover load resistance is three, as it should be for this ratio.

The energy balance is correct. The output shaft load force of one, is moving at three times the input shaft speed. The energy demand to drive the output shaft is three load forces. The input shaft speed of one is carrying one mechanical load force, and two hydraulic load forces, for a total of three. The input shaft is supplying the three load forces of energy that the output shaft needs to be driven at the 1:3 ratio.

Schematic 5 is further evidence that the purpose of the Third Law reaction forces is to provide the data and the mechanics for the Fourth Law to be able to calculate and extract the correct energy from the prime mover for the transmission of power. The power supplied is in alignment with the actual power required for driving automotive transportation.

Hydra-Mechanical Transmission at Infinity Ratio
Schematic 6

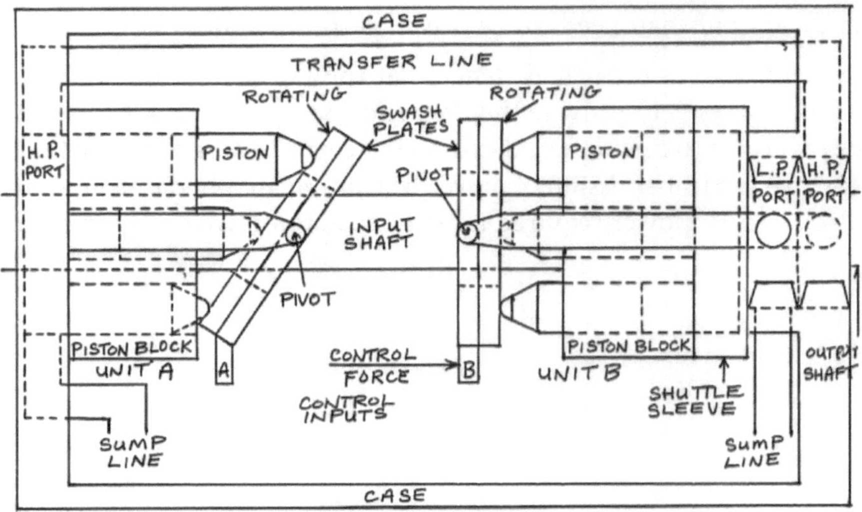

Schematic 6 illustrates the highest possible ratio that can exist between an input shaft and an output shaft, the infinity ratio. This ratio, the highest of the entire range of ratios, is only possible with a moving output shaft, and a non-moving input shaft. The moving output shaft is moving at a designated speed, not at infinity speed, and its relationship with the non-moving input shaft determines the infinity ratio. Mathematically, the infinity ratio is any ratio where the dividend of a fraction is a defined quantity and the divisor is zero. This is the case, with the Fourth Law transmission illustrated above, at the top of the range of ratios between two shafts.

The symbolic representation of the input/output ratio breaks down at this upper limit of the range of ratios. All previous lower ratios represented the input shaft speed with a "1." This representation helped to understand ratios throughout the entire range of ratios, with a designated speed for the input shaft. The implication of the "1" symbolism was that the input shaft speed was unchanging, steady speed, and that the ratio change was progressive and homogenous as the output shaft increased in speed. But at the highest ratio of infinity, the input shaft speed is stationary and cannot be symbolically represented by a "1." It must be represented by its true speed of zero, "0." As the ratios approach infinity, the "1 to something" may be quite high such as, "1 to 200, 500, 1000," but the upper limit of the ratios requires that the input shaft is stalled at zero for complete compliance with the Fourth Law of motion and the law of energy.

Transmission operations remain similar to previous lower ratios. The input shaft enters into the transmission case from the left, and attaches to piston blocks A and B. Input shaft revs at this ratio are stationary, and no clockwise motion is possible. The output shaft, exiting the case at the right end is attached to the shuttle sleeve, pivot braces and pivots, and unit B swash plate. These units move together as one, and at lower ratios, apply forces to the output shaft and receive the reaction load forces returning from the output. But at this ratio, output force application is not possible. Control forces at tab A stabilize swash plate A at maximum angle. The control forces applied to tab B have moved swash plate B angle perpendicular to the input shaft and case. In this position, unit B pistons do not stroke in or out of the cylinders, and all oil flow is deadheaded. Any possible oil flow from unit A is blocked, by the zero-volume unit B. Depending on prime mover, hydraulic oil pressure may be at maximum.

Energy flow through the transmission at this ratio is not possible. The input shaft is non-moving. It is impossible for a non-moving shaft to be able to supply energy regardless of the torque it applies to the system. The deadheaded oil at block B stops pistons at block A from reciprocating. These non-stroking pistons transfer a clockwise lateral torque to swash plate A. The case then carries these forces by reacting against them with counter clockwise forces that stop prime mover motion, blocking any mechanical energy flow. Prime mover/case action/ reaction forces are equal. Input is at full force with no displacement or motion, resulting in no work energy transfer. Unit B, at the square swash plate angle, would not be able to generate any lateral hydraulic forces that could be applied to power the output shaft. No energy could be applied to the output shaft with swash plate B in this position. Output is at full displacement, freewheel speed, with no applied force, resulting in no work energy transmission. In this ratio, both Unit A and unit B, individually and collectively, are incapable of performing mechanical work or the transmission of power.

This infinity ratio of the transmission would have only one useful application for automotive transportation purposes. Full inertial braking and acceleration would be possible, with this transmission. Design drivelines would vary, but a separate transmission responsible for braking only would have the input shaft connected to the wheels of the vehicle, and the output shaft connected to flywheels. During vehicle motion this "braking" transmission would be at the control ratio of the wheel speed to flywheel speed. When braking was required, control forces on tabs A and B would increase the wheel/flywheel speed ratio toward the highest infinity ratio. This process would slow wheel speed until locked stationary, at full stop, and increase flywheel speed, that are free-wheeling, until containing all of the inertial energy, originally contained in the moving mass of the vehicle. Then the energy from the flywheels could be returned by the drive transmission to the drive wheels, accelerating and driving vehicle mass, using stored inertial energy from the flywheels, without prime mover energy use, until needed.

The summarizing conclusion of Fourth Law transmission schematics follows in the next document.

TECHNOLOGY CONCLUSION

The manifestation of the Fourth Law of Motion is illustrated in schematic drawings above. Any transmission mechanism capable of the entire range of speed ratios between an input shaft and an output shaft, verifies the efficacy of the Fourth Law. Stated differently, no fully variable, full range transmission is possible without compliance with, and the existence of the Fourth Law of Motion. Implicit in the Fourth Law of Motion is the mechanical law of energy, which states that, work equals force multiplied by distance. Consequently, the schematics above show that the hydra-mechanical transmission, illustrated, is in compliance with the physics of energy and motion.

The compliant characteristics of the transmission are only possible with the proper application of hydraulic and mechanical power. This must take place, consistently, throughout the range of ratios. The direction of oil flow in the lower half of ratios, from "1 to 0" to "1 to 1," begins at unit B and is pumped to unit A, and does not change due to ratio change. The clockwise, downward moving B piston block is reciprocating pistons out of the block, due to swash plate angle, and is drawing oil from the sump into the block, in the near side of center. As block B continues to rotate upwards, on the far side of center, the pistons stroke into the block, due to swash plate angle and pump the oil out of the block, into the transfer line to unit A. Oil suction takes place from sump at block B, at front of centerline, and oil discharge back to sump takes place at far side of block A centerline throughout the entire lower range of ratios, from "1 to 0" to "1 to 1."

This consistency of oil flow orientation is maintained by the shuttle sleeve. The B swash plate, pivots, pivot supports, shuttle sleeve and output shaft move clockwise as a unit. As this "cluster" speeds up with increasing ratios, its relative motion to block B and the transmission case is constantly changing. The rotating shuttle sleeve constantly orients receding, sucking pistons to the sump, so that function remains the same, regardless of its position relative to block or case. Also, it orients pushing, pumping pistons of B block with the transfer line to block A, regardless of its relative position to block or case. The rotating shuttle sleeve maintains "near side of center/far side of center" hydraulic oil flow orientation and function, even though the entire cluster changes block B orientation and rotates with the output shaft.

Unit A receives process hydraulic oil flow from the transfer line and unit B. The A block clockwise downward motion reciprocates pistons out of the block, due to unit A swash plate angle. This makes the near side of center capable of receiving high pressure process oil, driving block A downward, assisting in input shaft rotation. Then as block A continues to rotate upward on the far side of center, the pistons reciprocate into the block, pushing out their oil, back to the sump, for use later in the transmission process. The swash plate, pivots and pivot braces do not rotate, as in block B, and are attached to the case. Therefore, the "near side of center/far side of center" function does not change at block A , and needs no shuttle sleeve to maintain component orientation.

"Near side/far side" piston block hydraulic function, and consistency of orientation has been established for the lower half of the range of shaft ratios. Near side block B and far side block A, are connected to oil sump, with unit B sucking oil from sump and unit A discharging to sump, with clockwise block rotation. Far side block B, pumps hydraulic transmission process oil to near side block A process oil, for the transfer of energy and motion with the same clockwise rotation. The orientation of these "near side/far side" functions, with clockwise shaft rotations, remains the same throughout all of the ratios, in the lower half of the range of ratios.

The same consistency of hydraulic function applies to the upper range of shaft ratios from 1 to 1 to infinity, 0 to 1, for compliant transmission function to take place. The relative motion of the B unit swash plate cluster and block B is reversed from the lower range of ratios. The output cluster is moving clockwise, near side down, faster than unit B piston block. As a result, the direction of transmission hydraulic oil flow is reversed, now flowing from unit A to unit B. Schematic 4 at the "1 to 2" ratio helps illustrate this consistency and reverse hydraulic oil flow and function.

Hydraulic "near side/far side" operations need to remain consistent at all high, upper half ratios. At 1 to 2, ratio, far side block A pistons, with upward motion, are reciprocating out of the block and drawing in oil from the sump, due to swash plate A angle. As the block continues to rotate, the near side downward motion reciprocates the pistons into the block, pushing hydraulic oil out into the transfer line to unit B. As the far side of output cluster B moves up, far side pistons reciprocate out of block B, able to receive transmission process hydraulic oil pumped from unit A. As output cluster of B continues to rotate, the near side pistons reciprocate into the block, pushing hydraulic oil back to the sump. This may be easier to visualize if we remember that the relative motion of block B is slower than the output cluster, during the upper half of the range of ratios. The same piston motion can be more easily visualized with a stationary B output cluster and a counter clockwise B piston block motion. The "near side/far side" consistency in oil flow port routing is maintained by unit B shuttle sleeve. Any further ratio changes in the upper range of ratios, only slows process hydraulic oil flow without changing "near side/far side" orientation to sump oil or transmission hydraulic process oil.

Operations integrity of hydraulic transmission process oil is maintained throughout the entire range of ratios of the transmissions drawn in the schematics. Even though hydraulic oil flow reverses direction half way through the range of ratios, pumping orientation remains the same throughout. Assuming clockwise rotation, the far side of unit A and the near side of unit B are always oriented to the hydraulic sump. Also, assuming clockwise rotation, the near side of unit A, and the far side of unit B are always oriented to transmission process hydraulic oil flow. The oil flow porting in the shuttle sleeve ensures hydraulic port orientation stability, even though unit B, sleeve cluster relative motion with unit B piston block and transmission case is always changing.

This stability of hydraulic oil operations is required for the transmission operations to be compliant with the Fourth Law of Motion and the machine law of energy. The schematics demonstrate transmission compliance access to the full range of ratios. The lowest ratio of "1 to 0," provides access to the ratio-based application of drive force to the mass of the automobile, without the use of energy from the prime mover. All vehicle speed change takes place as a result of homogenous ratio change, by controls sent to the transmission, without any prime mover speed change. The operating prime mover supplies the proper energy for all vehicle motion and applied force. All vehicle speeds are accessible by transmission ratio change only, without any prime mover speed change. All reaction forces returning from the driven output shaft are carried by B hydraulic unit. From there, the mechanical and the hydraulic energy flows are transmitted back to hydraulic unit A where the correct energy needed for the output drive is extracted from the prime mover. Reaction forces are routed to moving and fixed bodies as is required for energy demand and reaction force placement, all in compliance with the Fourth Law.

The science of the Fourth Law, and the technology of the schematics enables the alignment of automotive fuel use with the actual energy requirements for the acceleration and drive of the mass of the vehicle. **This is the fundamental correction of systemic automotive fuel use non-compliance, much needed in today's industry.** Initially, this was the entire motivation of this paper. But in the process, two unexpected, corollary technical characteristics became obvious. The technology of the hydra-mechanical transmission, makes these added advances possible. They are discussed in the following documents.

PART FOUR

Unexpected Developments

Introduction

The original motivation for the development of the science and technology of this paper was to correct automotive fuel misuse during the acceleration phase of vehicle motion. Graphs A, B, and C in the first problem section illustrate the initial inspiration for the work. But during the work in the development of the technology of the transmission, it became clear that advances could be made, not only in the acceleration phase of vehicle speed, but also in the steady speed and braking conditions of vehicle motion.

The characteristics of the transmission technology provide access to two advances of vehicle motion that could be beneficial. One benefit improves fuel use efficiency during steady state vehicle motion. The second advance improves vehicle energy use efficiency during vehicle braking, the deceleration phase of vehicle motion and vehicle speed-up, the acceleration phase of vehicle motion. The transmission characteristic that provides access to these advances is its totally homogeneous, variable ratio, complete range of ratios between two shafts, input and output. The specific traits of the transmission that contribute to the advances are, the variability of input/output shaft ratios, and the totality of the range of input/output ratios.

The homogenous variability of input/output shaft ratios, can be applied to all vehicle speed change. The vehicle speed controls are sent to the transmission. Then as the transmission ratio increases, the vehicle speed increases, but the engine speed does not need to change. This means that the engine provides only the energy needed for transportation, and does so at a steady engine rev speed. Therefore, steady speed engine design can be optimized for maximum fuel efficiency. The engine does not provide vehicle speed change by increasing its revs to speed up the automobile. The engine or motor is freed from providing speed change. And since it provides energy only, the engine can be engineered to be optimized for maximum energy efficiency at a steady design speed. The resulting optimized engine would greatly reduce steady speed vehicle fuel use. No longer would our steady speed, inefficient fuel use be enslaved to mediocre engine design due to prime mover speed change requirements.

The totality of the range of ratios of the transmission, provides access to the second unexpected advance in vehicle energy use. Vehicle braking with total inertial energy storage in flywheels would be possible. The transmission has access to the infinity ratio. The gradual contiguous ramping up to the 0 to 1 infinity ratio, with transmission input shaft speed at zero, stationary speed, and output shaft speed at free wheel, the infinity ratio, could be useful in vehicle braking. The access to the infinity ratio would drive flywheels with stored inertial energy from the braked or slowing vehicle. Then, when braking was finished, stored flywheel energy would accelerate vehicle mass when acceleration needed to begin. Stored flywheel energy from braking would be used for acceleration, without any engine energy demand, reduced and improved, through optimization, though engines may be.

The third and fourth potemtial developments are more systemic in nature than the first two, engine optimization and inertial braking. The third development could be the use of optimized engines for the generation of electricity and the possible threat to electrical grid operations. The fourth unexpected development could be the impact of the reduced atmospheric carbon di-oxide supply on forestation due to the automotive Fourth Law transformation. These potential developments are less directly related to the Fourth Law but still worth noting.

Engine optimization and total inertial braking and acceleration are the surprise, ontological advance benefits of this science and technology. These advances are clear and easily accessible given the science and the technology included in this paper. Also, there may be developments related to the thesis, but they would be more obscure and systemic in nature. Electrical grid operations may be impacted with the availability of optimized power plants at mass production prices. The health of forests may have to be monitored due to atmospheric changes. These are outlined in the documents that follow.

Internal Combustion Engine Optimization

The optimization of internal combustion engine fuel efficiency in the auto industry has not been achieved. This status-quo in design does not refer to the automotive misuse of fuel for the application of drive force at the beginning of vehicle acceleration as discussed in the problem section of this paper. This optimization refers specifically to internal combustion engine design. And the purpose of the optimization is to apply all design changes to the engine mechanism so that the maximum amount of mechanical energy can be supplied by the engine with the least amount of fuel use. These design changes are specifically for optimization of fuel use efficiency of the engine and not for engine race performance improvements. Race engine performance improvements have always been needed in the context of automotive fuel use non-compliance, and have been achieved at the cost of efficiency. But Fourth Law driveline compliance is now possible, and in this context, fuel use efficiency takes priority over power and race engine performance.

The barrier to internal combustion engine optimization in today's automotive state of the art is the engine's need to provide automotive speed change. For the vehicle to accelerate, the engine speed has to repeatedly increase from slow rev speed to a higher rev speed. The fixed ratio transmissions impose this operational requirement on the engines. Having to operate at high rev speeds, imposes a limit to the design compression ratio of the engine. Engines designed to increase revs to, say, 6000 revs per minute, are limited to 10 to 1 compression ratio. This is dictated by the combustion process in the combustion chamber above the piston. The stability of the combustion process in the chamber is enhanced by the octane fuel specification of the gasoline. Given the octane quality of commonly available gasoline, the compression ratio cannot be higher than the nominal 10 to 1 ratio if the engine is to rev up to the 6000 revs whenever needed. Any higher compression ratio at the higher rev speed would result in "knocking," or unstable, potentially destructive "autoignition" combustion in the combustion chamber, potentially damaging the engine.

The essence of the internal combustion engine efficiency limitation is the low compression ratio. The efficiency of the internal combustion engine increases with higher compression ratios. Typically, most automotive prime movers use the "Otto," four stroke combustion cycle.

The well-known, Otto cycle, efficiency equation confirms the relationship between increased engine compression ratio and increased engine fuel efficiency. Within limits, the higher the compression ratio, the higher the engine fuel use efficiency. The higher fuel use efficiency reduces engine heat loss out of the engine exhaust system and engine cooling system, and converts more of the energy of the fuel into mechanical energy, out the crankshaft. The higher efficiency engine wastes less heat out exhaust and cooling losses and produces more mechanical energy out the crankshaft for useful work, such as vehicle motion. The result is increased vehicle displacement motion, with less fuel use to produce that motion.

Two characteristics of high compression combustion lead to reduced heat losses and improved mechanical power production. The first, the stoichiometric burn temperature of the combustion is higher. The expansion ratio of the longer stroke, is the second. The higher burn temperature of the combustion for the stoichiometric fuel/air chemical reaction, requires less fuel to maintain the chamber combustion process temperature. Then when the heat has been generated in the chamber, using less fuel, more of that heat is converted to mechanical energy at the crankshaft due to the higher expansion ratio of the high compression ratio engine design. The resulting lower exhaust gas temperature reduces heat losses to the atmosphere. The lower the heat loses out the exhaust and cooling systems and the higher the transfer of energy to mechanical crankshaft work, the lower the fuel use for the actual work done. The result is the high efficiency engine. **These general principles of internal combustion engine optimization have been researched and achieved by naval engineering in Scandinavia and others.**

Engine speed is reduced to the lowest practical operational level. The compression ratio of the combustion chamber is increased as high as possible for steady, sustainable operations given the lower engine speed. Intake air supply is never throttled, thereby maintaining the high compression ratio. Engine throttling is achieved by the control of fuel to the combustion chamber. Fuel is introduced to the combustion chamber by sequential direct injection nozzles. The proper combustion of varying fuel air ratios due to throttled engine loadings is managed by the control of the injection nozzles. In so doing, the expansion ratio of the high compression ratio engine, stays at maximum, even though engine outputs are throttled. Engine volumetric size would be downsized depending on the torque and power needed for steady state vehicle operations. Cooling system and exhaust system designs would have to be downsized for the reduced heat loadings. Using these principles, naval prime mover efficiencies approaching 75%, a marked improvement over today's automotive benchmark of 25%, are all possible due to the steady speed engine design.

Total power output of the automotive optimized engine would be much less than the engines of today. Todays automotive engine is sized according the zero-efficiency force generation for the application of drive force for the acceleration of the mass of the vehicle. Therefore, it needs to be much higher and larger than steady state operational requirements. But the optimized engine and hydra-mechanical transmission need only apply the correct and much smaller amount of energy actually needed to move the mass of the vehicle. The operating optimized engine, idling at minimum fuel use would not be needed to provide massive amounts of power to begin the acceleration of the vehicle, as with the vehicles of today. The transmission outlined in the schematics would apply full motive forces for vehicle acceleration without any loading of the engine. Acceleration could be very forceful and engine energy demand would only be what was actually required for the forced motion of the vehicle mass, according to the mechanical law of energy. In so doing, engine power sizing would be much reduced from the automotive engine sizing standard of today, without a reduction in transportation performance.

Engines large enough for naval shipping make access to optimization easier than smaller automotive sized power plants. But given the principles of optimization and the downsizing benefits available due to the compliant schematic transmission, **significant advances to automotive fuel use compliance are possible.** This is the first of the unexpected advances that became evident during the development work of the Fourth Law transmission. The second, in the following document, is full inertial breaking with flywheel energy storage.

Unexpected Developments
Flywheel Inertial Braking

The understanding of the Fourth Law of Motion and its implications in the real world, lead to the development of the compliant hydra-mechanical transmission outlined in the schematics. It means that any transmission that is compliant to the Fourth Law, has to have access to the complete range of possible ratios, from 1:0 to 0:1, between an input shaft and an output shaft of a transmission. Conversely, the existence of any transmission capable of the full range of shaft ratios, from 1:0 to 0:1, presupposes the existence of the Fourth Law of Motion. The lowest ratio, 1 to 0, and the highest ratio, 0 to 1, have to be present and accessible to verify the efficacy of both the law and the transmission. Without access to these ratios, the lowest ratio and the highest infinity ratio, the possibility of inertial braking would disappear.

The infinity ratio makes inertial braking entirely possible. The relationship of the shaft speeds at infinity ratio is 0:1, zero to one. The zero speed of the input shaft is the actual motion speed of the shaft. At infinity ratio, it is not moving. The "1" of the output shaft is the symbolic representation of the shaft speed at whatever motion speed it is, at 10 rpm or 10,000 rpm, or any other speed. The infinity nature of the ratio is determined by the "zero" speed of the input shaft, a relationship defined by mathematics. Any "zero" speed of the input shaft requires an action/reaction relationship with the fixed body transmission case to verify and obey the Fourth Law, and such is the case with the transmission in the schematics. This characteristic of the infinity ratio makes full vehicle braking, with "zero" wheel speed possible. The infinity ratio "1" speed of the output shaft potentially driving the flywheels would allow any resulting flywheel speed according to the amount of stored inertial energy from the vehicle braking process. Regardless of flywheel speed, the ratio of the input to output shafts would be defined as infinity as long as input shaft speed was "zero," at the end of the braking process. And at that ratio, no further energy could be transferred to the flywheels, either mechanically by transmission, or theoretically, according to the Fourth Law of Motion.

The other characteristic of transmission operations needed for flywheel inertial braking, would be the smooth homogeneous access to the infinity ratio from the lower ratios. With the vehicle wheels in motion and flywheel speed at nominal, the input/output ratio would be less than infinity and at some place in the range of ratios of the transmission. Suppose that this ratio was 1 to 3, and the braking process started. Transmission control forces would move control swash plates so that the ratio would increase, totally variably, to ratios higher than 1 to 3, removing energy and slowing vehicle wheels, and adding energy and speeding up the flywheels. This process would continue until vehicle wheel speed stopped at zero and nominal flywheel speed stored vehicle inertial energy, at the infinity ratio.

Then, when braking is over and vehicle speed control is returned back to acceleration, the flow of energy between wheels and flywheels would be reversed. The flywheels, at nominal speed, would be moving and be the source of power. The vehicle wheels would not be moving and would be receiving the energy needed for acceleration. The moving flywheels, driving the non-moving vehicle wheels is the 1 to 0 ratio, the lowest transmission ratio, and at the opposite end of the infinity ratio, required for inertial braking. Reversing the energy flow from braking to acceleration, moves the control operating ratio of the transmission from the highest ratio, 0 to 1, infinity, to the lowest ratio, 1 to 0. Inertial braking and acceleration require full ratio range dynamics, only possible with the Fourth Law and the Fourth Law compliant transmission.

The potential fuel consumption savings are truly transformational. The flywheel inertial braking is two full design systems of efficiency improvement over and above today's automotive state of the art. The first design improvement is the Fourth Law and the hydra-mechanical technology that lowers automobile fuel use from maximum fuel volume, wasted at zero efficiency, down to fuel use alignment with the actual energy requirements for the forcing, acceleration and drive of the vehicle. This eliminates all automotive fuel use waste that takes place in today's transportation merely for the application of the drive force. Even though the waste is eliminated, this design system still requires the correct amount of fuel for the energy needed to accelerate the mass of any vehicle. But this second flywheel design system eliminates even this "correct" driveline fuel use during the acceleration phase of the vehicle. The inertial braking and acceleration flywheel design system not only eliminates the wasted fuel use for vehicle acceleration of today's automotive standards, it also eliminates ANY driveline fuel use for the acceleration of the vehicle. It improves actual, correct fuel use to no fuel use during the acceleration of the vehicle. Only when flywheel energy is exhausted due to the acceleration of the vehicle is fuel use needed by the driveline for the further transportation of the vehicle, at actual and correct volumes.

The Fourth Law science and the related hydra-mechanical transmission technology has the potential of initiating positive transformational changes. These massive changes may have the potential of less obvious, systemic effects that should be considered. These larger systemic effects could be considered desirable, and are explored in the following documents.

Related Systemic Obstructions

Given the monumental nature of the proposed transformation, massive impacts or related potential obstructions are inevitable. Any systemic impacts related to the thesis would have to be a result of the manifestation, or the physical presence of the Fourth Law of Motion in the real world. The technology that makes the transformation possible, would be the source of the changes, even at this systemic level.

Suppose the transformation takes place. What would be the single most significant change in the physical reality of our existence? The technology of the transmission in the schematics would allow and demand the optimization of the internal combustion engine. With engines available at efficiencies of, say, 65 to 70 percent, at automotive mass production prices, would other uses for these engines be found? Considering the cost of electricity, with its mediocre grid thermodynamic efficiency, and the supply of transformational, gensets for home service, the generation of electricity could become localized. House service high efficiency engines, designed for automotive use, would be available that could drive generators for less cost, both financial and environmental due to the increased efficiency. And larger sized high-efficiency engines, designed for automotive use, could drive generators capable of powering multiple homes, from several dwellings to community sized service, possibly for less cost, with higher thermodynamic efficiency and the reduced carbon footprint than the less efficient grid. It could be that the presence of the readily available, high efficiency internal combustion engines, would call into question the justification of the lower efficiency operation of the grid. Such a disruption could only lead to improvements of the grid that would be beneficial for all, including an improved carbon footprint. But the transformation from the old to the new may be difficult.

More obscure than the systemic negative impact on the electrical grid, would be the disruptions of reduced carbon supply, released to the environment. The automotive supply of millions of tons of carbon di-oxide, released to the environment every day, would be drastically reduced when vehicle fuel use efficiencies improved to 65% or higher. Consider the impact of this radical reduction of "tree food" on the health and vigor of the forests. Trees, that are constantly adapting and evolving to and in the environment of ever-increasing carbon "tree food" due to low automotive fuel use efficiency, may suffer when this rich carbon supply ends. Tree photosynthesis, using carbon di-oxide from the air, and energy from the sun supplies trees with the energy and nutrition needed for growth. When the daily millions of tons of this rich "tree food" is reduced, trees may suffer from lower atmospheric carbon loadings. The life cycle of most trees is longer than the rise and fall cycle of carbon loading. The life span of trees in rising carbon loadings encourages tree vigor and health in an increasingly rich environment. The reduced carbon loadings and falling "tree food" supply, would force these same trees to adapt to the more difficult environment of reduced "tree food" supply. The concern is that trees that have evolved and adapted to the easier environment of increasing food supply, may suffer in the more difficult environment of less atmospheric carbon-based food. Are the trees, accustomed to easier, higher "tree food" supply, capable of the more difficult adaptations when atmospheric carbon loadings are reduced due to the automotive efficiency transformation? Would the foliage of the forests be less vigorous as a result? The environment of steadily reducing "tree food" supply could result in tree foliage changes that we should be aware of, even though the reduced carbon release would be beneficial in a global sense.

Obstructions other than the electrical grid and forestation, exist, without doubt. The management of hydrocarbon resources would be heavily impacted if the transformation to responsible automotive fuel use, as per this paper, took place. No doubt business economics and the principles of supply and demand, would steer the course through the changing environment.

Unexpected Developments
Summary

The initial purpose of the transformation outlined in the paper was to reduce automotive fuel misuse related to the non-compliance of the application of motive force to accelerate and drive vehicle mass overall. The purpose of the research was to eliminate the waste of automotive fuel use. The relevant science and technology were generated to bring about compliance to the laws of energy, and resulted in the solution to the fuel misuse problem. During the research process for the development of the transmission technology, unexpected related changes became evident, both advantageous and systemic.

For the transmission to be compliant to the Fourth Law of motion, it had to access the full range of ratios that are possible between two shafts, an input shaft and an output shaft. This ability allowed the prime mover engine to operate at a single speed, and allowed vehicle speed change to take place, by way of the ratio change of the transmission. This steady speed engine, the first in the world of variable speed automotive engines, can be optimized for the maximum efficiency of fuel use. Research for marine engine efficiency has developed ship engines, approaching 75% fuel efficiency, due to their steady speed operations. The same research for steady speed auto engines would have similar results improving efficiency from the current 20% auto engine efficiency, to at least 65 or 70% during steady speed highway operations. As a result, vehicle fuel mileage for highway use could double or more.

The second corollary surprise was the possibility of full inertial braking due to the transmission's access to the highest infinity ratio. The infinity ratio, 0 to 1, representing the relationship of the vehicle wheel speed and the speed of the energy storage flywheels, makes inertial braking possible. When fully stopped the vehicle wheels are at zero, "0," speed and the flywheels are at the resulting energy storage speed represented as "1." The other characteristic of the transmission that allows access to infinity ratio, is the fully homogenous access to every incremental ratio throughout the entire range of ratios. This allows access to the infinity ratio from whatever ratio of vehicle wheel speed and flywheel speed at the beginning of braking. As the braking starts, the wheel/flywheel ratio increases, the vehicle slows down and the flywheels speed up, taking in the inertial energy of the slowing vehicle. At the arrival of infinity ratio, the vehicle is stopped and the flywheel is at energy storage speed. Then when acceleration starts, the energy flow is reversed and so the transmission ratio drops from infinity to the lowest, beginning ratio of 1 to 0 for the return to acceleration mode. The inertial energy from the flywheels, accelerates the vehicle without the use of energy from the prime mover. Then when vehicle motion increases to highway speed, energy is supplied by the steady speed prime mover engine, at the suggested high efficiency, reduced fuel consumption rate.

In the wake of advances such as these, there could be systemic changes for which we should be prepared. High efficiency steady speed engines would be ideal for the generation of electricity at the local house sized application. Larger high efficiency community sized generation stations could be developed producing electricity at higher efficiencies than the grid. Depending on numbers, the grid may have a difficult time justifying its comparatively low efficiency existence, never before questioned in the absence of a high efficiency alternative. But if engines of, say, two thirds or 66% efficiency became available, at mass production prices, and average grid thermodynamic efficiencies remained at the typical 40%, the cost and the environmental impact might be enough motivation for localized, micro-electrical generation, perhaps larger in scope, threatening the status-quo existence of the grid.

Another systemic change could be the impact on forestation in the wake of the end of the daily release of millions of tons of carbon di-oxide into the atmosphere. The transformation would align automotive fuel use with the actual energy needed to force, accelerate and drive the mass of the vehicle, all vehicles. The fuel savings would be enormous. Automotive waste fuel use would be eliminated. Equally massive would be the reduction of carbon di-oxide released to the atmosphere. The past increasing levels of atmospheric carbon are well with in the capacity of photosynthesis of trees, and this added "tree food" has served the foliage of the forests well. The concern with the reduced carbon supply to the environment is the potential impact of this reduced "tree food" on the health of trees. The lower or steady carbon levels would remain in the range for tree photosynthesis, but would be more difficult to adapt to, than the increasing supply. Trees, acclimatized to easy, rising carbon adaptation, would be forced to the more difficult adaptation with the steady or reduced carbon supply. The goal would be that tree health would remain the same, or increase, in the environment of stable or steadily reduced carbon supply.

The presence of the Fourth Law of Motion would bring about many changes, transformational and beneficial. The conversion of automotive fuel use, away from non-compliance for the forcing, acceleration and drive of the mass of the vehicle, to compliance with the mechanical law of energy, is the transformational change brought about by the law. This aligns fuel use with the actual energy needed to move the mass of the vehicle, a transformation away from the present automotive use of fuel. Automotive fuel use waste would be eliminated. Subsequent beneficial enhancements, related to the transformation are possible. Research leading to highly efficient auto engines is now possible and ready to be applied to automotive transportation. The advance of full inertial braking would greatly improve vehicle acceleration fuel use efficiency. Systemic changes like the higher efficiency generation of electricity, would be of benefit to society at large. The unheard-of question of the impact of reduced carbon loading, of reduced "tree food" on trees and forestation, would have to be addressed, and could result in positive benefits. These would be the possible impacts of the Fourth Law. Other impacts no doubt exist and will be managed as they arise.

PART FIVE

SUMMARY

Many decades ago, while studying for the First-Class Stationary Engineering apprenticeship, I saw the automotive fuel use curve for the first time. The moment I understood, the high automotive fuel use at zero efficiency at zero vehicle speed, I realized the inconsistency with the High School physics equation for mechanical energy. The mechanical law of energy states that the application of force with zero displacement requires no energy. Yet the automotive fuel use curve showed maximum fuel use, at zero efficiency, at the beginning of automotive acceleration, upon application of force, but before motion had started, at zero speed. The realization was that automotive fuel use was totally in non-compliance with the work energy laws. The realization of the significance of the problem of automotive fuel use waste, provided the inspiration and motivation for the research and development for the solution to the problem. This paper is the record of the work.

THE PROBLEM

The problem of the noncompliant automotive fuel use had to be defined. Once it was clearly understood, both the non-sustainability of the fuel misuse condition, and the potential of an energy compliant resolution, defined current automotive state of the art as a problem. Current automotive fuel misuse is easily defined. The worst misuse takes place at the beginning of acceleration of the vehicle. At the point of the application of motive forces to accelerate the vehicle, the drive line uses the highest volume of fuel, fossil fuel or electricity, at the lowest efficiency of 0%, zero percent. Repeated at the risk of over statement, the highest rate of fuel use, at any speed, is used at the lowest possible efficiency of zero percent, just to apply the motive force, before any motion has taken place. And, theoretically, no fuel should be used to apply this motive force. This condition of highest volume fuel use, at zero percent efficiency, to apply a drive force that theoretically needs no energy to apply could, conscientiously, be considered a total waste of fuel! This the characteristic of every automotive driveline.

Graphs A and D illustrate this misuse. The problematic nature of this condition is reinforced by the science, the mechanical law of energy. It states that no work is done until the applied force is moved through a displacement. This law condemns automotive fuel use in the harshest of terms. It states that automotive fuel use, fossil fuel or electric, should be zero at the point of application of the drive force for the onset of acceleration. The science says that auto fuel use should be zero when in reality, it is actually the highest volume, at lowest zero efficiency. This clearly defines the automotive fuel use state of the art as a problem.

A more systemic reinforcement of automotive energy misuse as a problem lies in the understanding of the first law of thermodynamics. It states that energy cannot be created or destroyed, only changed from one form to another. This means that the energy to force, accelerate and drive a vehicle is based on science and is always the same for the vehicle, given its mass and acceleration. This means that there is an actual, definable energy use for the acceleration of a vehicle mass and as such provides the target for which automotive research and development should strive. **This development has not taken place,** and the standard automotive fuel misuse takes place at every automotive acceleration, regardless of the actual energy required by the vehicle. Due to the mechanical energy definition of zero vehicle fuel use for the application of a force, today's automotive fuel misuse is converted to low grade heat and carbon products of combustion, all exhausted into the atmosphere. Exhausted needlessly into the atmosphere is the problem that could be resolved with alignment of automotive fuel use with actual vehicle energy requirements. Both heat and carbon have detrimental global environmental consequences when the massive, one and a half billion vehicles, world-wide, operate according to the same principles of non-compliance.

The possibility of the science-based resolution to the current energy misuse condition, further reinforces the statis quo as a problem. Given the massive automotive engineering research and development resources, in concert with the total lack of development in solving the noncompliance problem, suggests only one thing, that the science needed for the resolution of the problem does not exist. Systemic resolution of the problem can only take place with the discovery of new relevant science, in the form of the laws of motion. The new laws could only be relevant, if they provided the definition of the correct use of energy in the context of the motion of a forced mass or body. This science would lead to the technology that could be applied for the compliance of auto fuel use, and provide automotive research and development the resources needed to achieve compliant fuel use for correct operations.

The Science

The material missing from the physics that is needed for the understanding of compliant fuel use is the laws of motion that introduce and define energy use when a mass or body is driven by a force and begins to move. It is defined as the Fourth Law of Motion, since Newton's first three laws of motion are the point of departure for this next, relevant law. Newton's first two laws of motion define force, mass and acceleration in both qualitative and quantitative terms. Newton's third law defines the relationship between the forced body or mass and the source of the force applied to the body. It states that, "Every action has an equal and opposite reaction." In common terms this means that the body or mass pushes back on the forcer equal to the force that is pushing on the body. It is this Newtonian third law of motion that provides the point of departure to the next relevant, energy compliant law, labelled the Fourth Law of Motion.

The requirements of the Fourth Law are very specific. It has to get the energy right at all speed ratios between the source of force and energy, and the forced moving mass or body. The correct energy is defined by the mechanical law of energy, and it provides the context for energy use of a forced, moving body at all speeds, relative to the source of energy. In order to do this, the point of departure of the Fourth Law must be the "equal and opposite reactions" of the Third Law of Motion. These "equal and opposite reactions," the communications of the forced and moving body back to the source of energy, must be borne by the correct frames of reference through out the speed range of the moving body. These frames of reference extract the correct level of energy from the source of energy so that the forced, moving body is powered with energy according to the laws of thermodynamics. To do this, the Fourth Law of Motion states:

Equal and opposite reactions, act upon fixed bodies, moving bodies, or proportionally both fixed and moving bodies.

The essence of the breakthrough of the Fourth Law is its definition of two relevant frames of reference, the fixed body and the moving body. The existence of each is required for the science-based introduction of energy to a forced body that is beginning to move. Each frame of reference must have its own reaction force management system. These Fourth Law requirements enable reaction forces from the load to be managed in compliance with energy laws. Compliance requires that the reaction forces be shifted to and applied onto either the fixed body, or the moving body, or be shared by both the fixed and moving bodies. The shared reaction forces either compliment or provide reaction resistance forces between both fixed and moving bodies, given the relative speeds of the source of energy and the moving load. These are all of the possible reference body interactive relationships needed to extract the correct compliant energy for a moving forced load throughout its speed change and throughout the range of transmission shaft speed ratios.

The process that this Fourth Law initiates is the compliant use of energy for the driving of a forced and moving body, a vehicle. The reaction forces of the forced but stationary body are borne by a non-moving, fixed frame of reference. This requires no energy from the source in order to apply the drive force to the vehicle. This is in compliance with the laws. Then, as the body begins to move, the reaction forces begin to shift away from the fixed frame of reference onto the moving frame of reference, which is the source of energy needed to move the forced body or vehicle. Two frames of reference, or bodies, one fixed and one moving, are required by the Fourth Law of Motion. At these initial levels of energy extraction, the reaction forces must be borne, shared, appropriately, proportionally by both the fixed and moving frames of reference. Then as the speed of the forced moving body increases, the energy requirement will increase so that all of the reaction force will be borne by the moving frame of reference, the engine or source of energy, and no reaction forces will be borne by the fixed frame of reference. Technically this will be defined as the "one to one" ratio, as source torque and load torque will be the same. The fixed body caries no reaction forces at this point.

But the interaction process of the two frames of reference continues as the speed of the driven and moving body increases. The increase of energy, above 1 to 1, can now only be extracted with a proportional action/reaction relationship between the fixed and moving frames of reference. The higher torque loadings needed to extract higher levels of energy from the engine, the moving frame of reference, now have to be borne by the fixed frame of reference. The torque loadings above the load reaction forces need to be borne by the fixed frame of reference, so that the appropriate energy for driving the load can be extracted from the engine moving frame of reference. The fixed frame of reference provides the resistance needed to load the engine, so that it can supply the power needed to drive the load at speeds higher than the 1 to 1 ratio. This process of action/reaction continues under the rule of the Fourth Law, until the highest speed of the driven body is reached. At this infinity speed, the action/reaction relationship of the fixed and moving bodies is one to one. This means that the moving body is stalled at zero speed, applying full reaction forces on the normally fixed frame of reference, which is providing full reaction forces, to the moving body, now stationary. This action/reaction force relationship, between the fixed and moving frames of reference takes place at highest driven body or load speed, at "infinity ratio" speed. And under the rule of the Fourth Law, this means that no energy or force can be applied to the forced, driven and moving load. It is now freewheeling.

The Fourth Law of Motion defines the appropriate energy extraction needed for the driving of a forced mass or body. It does so, qualitatively, according the mechanical law of energy. Therefore, it is suggested that the quantitative Fifth Law of Motion should be the mechanical law of energy:

Work equals Force times Distance. (W=F x D)

This equation states that both force and distance or displacement must have a value for "work" to have a numerical quantity. If either "force" or "displacement" are zero in quantity, then "work" or the related energy of the work is also zero. This relationship clearly states that the application of force without a distance does no work, and requires no energy. It describes the essence of the non-compliant fuel use of the automotive industry today.

The understanding of the interplay of the Fourth and Fifth Laws is important. The Fourth law is qualitative in nature and provides the physical mechanisms and material process needed for energy compliant based science of motion. The interaction of the fixed and moving frames of reference of the Fourth Law does this. But the quantitative Fifth Law of Motion, controls and imposes the principles of energy use onto the Fourth, so that every load speed, at every reference body interaction extracts compliant amounts of energy. The Fourth Law's obedience to the Fifth, guarantees energy use compliance, at all load speeds, vehicle or otherwise.

This is the science that provides the framework for any related technology that could advance auto fuel use into energy compliance in the real world. The technology that manifests the Fourth and Fifth Laws of Motion into the real world, follows.

The Technology

The technology that falls out of the Fourth and Fifth laws is a transmission. The transmission transforms the physics into the physical reality of the world. The invisible is made visible. The physics discovered the thermodynamics for compliant automotive fuel use and solved the auto fuel misuse problem. The physics compliant transmission provides the technology for the automotive world such that potential transportation fuel use is also in compliance with the solution to the problem provided by the physics. The ontology of the physics of the Fourth and Fifth Laws is compliance to the laws of thermodynamics that states that fuel use energy for a vehicle should, must, equal the energy needed to force, accelerate and drive the mass of the vehicle. These are the fundamentals of the physics of energy. It is the function of this compliant transmission to transform all auto fuel use into compliance with the Fourth and Fifth Laws of Motion and the laws of thermodynamics.

The physical presence of the transmission is similar to any other. It is a metal box located between the prime mover source of energy and the drive wheels of the vehicle, the load. Input and output shafts connect the transmission box to the engine and drive wheels or flywheels as per application. These shafts are connected to internal hydra-mechanical pumps capable of transmitting energy from source to load as needed. The controls for vehicle speed are received by the transmission and these controls are capable of power engagement and ratio changes, over the entire range of ratios. All controlled ratio changes would extract correct energy use amounts as defined by the energy laws.

The specifications of the transmission that operates in compliance with the physics are highly defined. The function of the transmission can be considered proof of the efficacy of the Fourth and Fifth Laws, due to its exact compliance. The range of ratios, from lowest to highest, must be all possible ratios, between an input shaft and an output shaft. The lowest ratio would be 1 to 0, with the moving input shaft, represented by "1," and zero speed output shaft for the drive wheels, represented by "0." The highest ratio would be 0 to 1, the infinity ratio, with the input shaft at zero speed and the moving output shaft speed represented by "1." Access to this entire range of ratios would also require equal access to every incremental change of ratio, throughout the entire range of ratios, from the lowest to the highest. Access to the full range of ratios, and every incremental ratio within the entire range of ratios is achieved by the control and position of the swash plates of the hydra-mechanical hydraulic pimps.

Mechanically, the transmission control input forces position the swash plates so that ratio access is complete. But mechanical manipulation of the ratios is only possible due to hydra-mechanical compliance to the theoretical requirements of the Fourth Law. Every incremental ratio change requires a placement of the reaction force of the vehicle load onto a fixed body, a moving body, or proportionally onto both. The transmission uses two energy management systems, one mechanical and one hydraulic, so that full access to reaction force placement, by the mechanical positioning of the swash plates and the vehicle control forces, is possible. This is necessary so that the amounts of energy extracted from the prime mover and applied to the load are in compliance with the Fourth and Fifth Laws of Motion and the all controlling laws of thermodynamics. The transmission's obedience to these laws guarantees the alignment of vehicle fuel use with the actual energy requirements to force, accelerate and drive the mass of the vehicle.

The transmission performs like a mechanical power transformer. The Fourth and Fifth laws require that the correct energy is extracted from the prime mover to drive the forced and moving vehicle. To do this the transmission takes power from the steady speed variable load engine shaft and applies it to the variable speed steady load force output shaft. To do this the transmission has to transform power from steady shaft speed to variable shaft speed and from variable supply load force to steady output load force. It has to make this energy transformation throughout the entire range of input/output ratio at every homogenous incremental ratio. This energy transformation process guarantees the correct use of energy as defined by the Fourth and Fifth Laws of energy and fuel use in compliance with the laws of energy. The resulting reduction in automotive fuel use and atmospheric carbon loading would be enormous and of great benefit. The specific impacts of the science and technology, could be far-reaching, beneficial and possibly, systemic.

Transformational Impacts

The impacts of the Fourth Law are transformational. In keeping with the most far-reaching of the responsibilities of human existence, the Fourth Law addresses a problem that heretofore has been irresolvable. Until now, automotive fuel use for the application of a drive force for the acceleration of a vehicle has been at maximum fuel volume, when theoretically it should be zero! The Fourth and Fifth Laws and the related compliant transmission technology make this fuel use correction possible. The correction of this fuel misuse problem is transformational in nature. It could only have been achieved with the development of the science of the laws of motion that govern the energy use by a forced and moving body, the Fourth and Fifth laws. These laws led to the obedient technology that complied to the laws, and solved the fuel misuse problem, a correcting solution, transformational in nature. The resolution of this problem was the original inspiration and the main purpose for the research and development for the science and technology, and the record of such in this paper.

During the research for the transmission technology, two unexpected energy related potential advancements were realized. These advancements are technical in nature and arise out of the design ratio range of the transmission. The total ratio range variability of the transmission can now be used to control power engagement and vehicle speed change, never before possible in automotive drive line operations. This transmission capacity would allow auto speed change to take place without a change in engine rev speed. With this transmission, auto prime movers would operate at a single design speed and could then be optimized for maximum fuel efficiency at that engine rev speed. The higher the engine compression ratio at that single engine rev speed, the higher the fuel efficiency of the engine, possibly three times the efficiency of existing variable speed auto engines of today.

The second technical advancement due to the totality of transmission ratio range is the potential development of full inertial braking of the vehicle. Inertial braking requires the highest possible ratio, the infinity ratio, for flywheel, energy storage braking, and then the lowest possible ratio to return the flywheel energy to the vehicle for acceleration. This Fourth Law transmission, has access to the infinity ratio and the lowest, 1 to 0, ratio, to make inertial braking possible. Braking would begin under transmission control force, at the vehicle wheel speed to flywheel speed ratio. Transmission control force would increase this ratio to slow the wheels and speed up the flywheels. Vehicle inertial energy would be stored in the flywheels. At the highest ratio of infinity, the wheels would be stopped and the flywheels would be freewheeling, containing all vehicle inertial energy. Then the transmission would shift to the lowest ratio of 1 to 0, with flywheel speed represented by 1 and vehicle wheel speed of 0. Transmission control forces would increase ratios from this lowest ratio to slow flywheel speed, removing their inertial energy and returning it back to the vehicle wheels for acceleration, without any energy use from the engine for acceleration.

The transformational vehicle fuel use alignment to thermodynamic, and the two technical transmission ratio range benefits, engine efficiency and inertial braking, can be considered advancements. Other systemic impacts related to the fourth law could also be present. The high efficiency auto engines could present a threat to the electrical grid if house hold electrical energy could be generated less expensively by high efficiency engines than on the grid. Also, the Fourth Law related reduction of automotive carbon exhaust may level or reduce the atmospheric "tree food" supply, to the detriment of the health of tree foliage, adapted to ever increasing supplies of carbon dioxide, as they are now.

Advancements and potential systemic impacts related to the new science of motion could be beneficial in the solution of the related problems.

CONCLUSION

Decades ago, the realization of the extravagant wastefulness of automotive fuel use, opened the awareness to any and all information related to this fuel misuse problem. Then the first sight and understanding of the graphs A, automotive fuel use, and D, automotive fuel use efficiency, inspired the research and development that led to this document. The graphs showed that automotive fuel use was in total non-compliance with $W = F \times D$, the correct theoretical fuel use for a forced and moving body. The work was on.

Two moments of inspiration, in decades of work, lead to the solution of the automotive fuel misuse problem. The discovery of the Fourth Law of Motion was one. The discovery of the compliant transmission design was the other. When these discoveries were fully understood, it was realized that the automotive fuel misuse problem could be solved, and that transportation energy use could be shifted into alignment with actual theoretical, vehicle fuel use requirements. Fuel consumption savings would be enormous. Reduction in carbon discharge loadings would be just as large and possibly even more significant. The reduced demand on hydrocarbon supply resources would lead to sustainability. The reduced, millions of tons of daily carbon release to the atmosphere would lead to global temperature and environmental sustainability. Resource development and the atmospheric environment would not have to continue to cushion and pay the price of the existence of the needs of human transportation individualization. Harmony, co-existence and peace, may become reality.